Contents Tourism and Pop Culture Fandom

ASPECTS OF TOURISM

Series Editors: Chris Cooper (*Leeds Beckett University, UK*), C. Michael Hall (*University of Canterbury, New Zealand*) and Dallen J. Timothy (*Arizona State University, USA*)

Aspects of Tourism is an innovative, multifaceted series, which comprises authoritative reference handbooks on global tourism regions, research volumes, texts and monographs. It is designed to provide readers with the latest thinking on tourism worldwide and in so doing will push back the frontiers of tourism knowledge. The series also introduces a new generation of international tourism authors writing on leading edge topics.

The volumes are authoritative, readable and user-friendly, providing accessible sources for further research. Books in the series are commissioned to probe the relationship between tourism and cognate subject areas such as strategy, development, retailing, sport and environmental studies. The publisher and series editors welcome proposals from writers with projects on the above topics.

All books in this series are externally peer-reviewed.

Full details of all the books in this series and of all our other publications can be found on http://www.channelviewpublications.com, or by writing to Channel View Publications, St Nicholas House, 31–34 High Street, Bristol BS1 2AW, UK.

ASPECTS OF TOURISM: 88

Contents Tourism and Pop Culture Fandom

Transnational Tourist Experiences

Edited by
Takayoshi Yamamura and Philip Seaton

CHANNEL VIEW PUBLICATIONS
Bristol • Blue Ridge Summit

DOI https://doi.org/10.21832/YAMAMU7222
Library of Congress Cataloging in Publication Data
A catalog record for this book is available from the Library of Congress.

Library of Congress Control Number: 2019029539

British Library Cataloguing in Publication Data
A catalogue entry for this book is available from the British Library.

ISBN-13: 978-1-84541-722-2 (hbk)
ISBN-13: 978-1-84541-721-5 (pbk)

Channel View Publications
UK: St Nicholas House, 31–34 High Street, Bristol, BS1 2AW, UK.
USA: NBN, Blue Ridge Summit, PA, USA.

Website: www.channelviewpublications.com
Twitter: Channel_View
Facebook: https://www.facebook.com/channelviewpublications
Blog: www.channelviewpublications.wordpress.com

Copyright © 2020 Takayoshi Yamamura, Philip Seaton and the authors of individual chapters.

All rights reserved. No part of this work may be reproduced in any form or by any means without permission in writing from the publisher.

The policy of Multilingual Matters/Channel View Publications is to use papers that are natural, renewable and recyclable products, made from wood grown in sustainable forests. In the manufacturing process of our books, and to further support our policy, preference is given to printers that have FSC and PEFC Chain of Custody certification. The FSC and/or PEFC logos will appear on those books where full certification has been granted to the printer concerned.

Typeset by Nova Techset Private Limited, Bengaluru and Chennai, India.
Printed and bound in the UK by Short Run Press Ltd.
Printed and bound in the US by NBN.

Contents

Contributors vii
Acknowledgements xi
Foreword xiii

Introduction: Contents Tourism Beyond Anime Tourism 1
Takayoshi Yamamura

Part 1: The Contentsization of Literary Worlds

1. The Contents Tourism of Jane Austen's American Fans 19
Philip Seaton

2. Conceptualizing Contents Brandscapes: The Brontë Brand 34
Maree Thyne and Gretchen Larsen

3. *The Witcher* Novels and Games-inspired Tourism in Poland 46
Aleksandra Jaworowicz-Zimny

4. Travelling *Heidi*: International Contents Tourism Induced by Japanese Anime 62
Takayoshi Yamamura

Part 2: Tourist Behaviours at 'Sacred Sites' of Contents Tourism

5. The Cotswolds and Children's Literature in Japanese Fantasy: The Case of Castle Combe 85
Catherine Butler

6. Yōkai Tourism in Japan and Taiwan 98
Shinobu Myoki

7. Contents Tourism and Religious Imagination 116
Kyungjae Jang

8. The 2.5-Dimensional Theatre as a Communication Site: Non-site-specific Theatre Tourism 128
Akiko Sugawa-Shimada

9. Indonesian Cosplay Tourism 144
Ranny Rastati

Part 3: Contents Tourism as Pilgrimage

10 Outbound Tourism Motivated by Domestic Films: Contentsized Koreanness in Thai Movies and Tourism to Korea 161
    Sueun Kim

11 Contents Tourism in Plane Sight 174
    Christopher P. Hood

12 Breaking Benjamin: A Woman's Pilgrimage to New Mexico 189
    Stefanie Benjamin

13 From Banjo to Basho: Poets, Contents and Tourism 205
    Sue Beeton

Conclusions: Sustainable Contents Tourism in the 21st Century 224
Philip Seaton

Index 244

Contributors

Editors

Philip Seaton is a Professor in the Institute of Japan Studies, Tokyo University of Foreign Studies. He researches war memories and contents tourism relating to Japanese history in the period 1853–1945. He is the author/editor of four books: *Japan's Contested War Memories* (Routledge, 2007), *Voices from the Shifting Russo-Japanese Border* (with Svetlana Paichadze, Routledge, 2015), *Local History and War Memories in Hokkaido* (Routledge, 2016) and *Contents Tourism in Japan* (with Takayoshi Yamamura, Akiko Sugawa-Shimada and Kyungjae Jang, Cambria Press, 2017). He has also guest edited special editions of *Japan Forum* and *Journal of War & Culture Studies*. His website is www.philipseaton.net.

Takayoshi Yamamura is a Professor in the Center for Advanced Tourism Studies, Hokkaido University and holds a PhD in urban engineering from the University of Tokyo. He is one of the pioneers of 'Contents Tourism' and 'Anime-induced Tourism' studies in Japan and he has served as the Chair of several governmental advisory boards such as the Meeting of International Tourism Promotion through Animation Contents of The Japan Tourism Agency, ANIME-Tourism Committee of Saitama Prefecture, etc. His main English work includes *Contents Tourism in Japan: Pilgrimages to 'Sacred Sites' of Popular Culture* (with P. Seaton, A. Sugawa-Shimada and K. Jang, 2017), *Japanese Popular Culture and Contents Tourism* (co-edited with P. Seaton, 2016), 'Pop culture contents and historical heritage' *Contemporary Japan* (30.2) and 'Contents tourism and local community response', *Japan Forum Special Edition* (27.1). http://yamamuratakayoshi.com/en/.

Authors

Sue Beeton is a travel and tourism researcher and writer and President of the Travel and Tourism Research Association (TTRA), Foundation Chair of the College of Eminent Professors at William Angliss Institute in Australia and an Associate at La Trobe University. For over a quarter of a century, she has conducted tourism-based research into community development, film-induced tourism and pop culture and nature-based tourism. As well as producing numerous academic papers, book chapters and reports, Professor Beeton has published a range of research-based books, including *Ecotourism: A Practical Guide for Rural Communities, Community Development Through Tourism* and *Tourism and the Moving Image*, as well as two editions of the acclaimed monograph, *Film-Induced Tourism*. She was recently awarded a Lifetime Achievement Award for her contribution to tourism research and scholarship.

Stefanie Benjamin, PhD, CHE is an Assistant Professor in the Retail, Hospitality and Tourism Management department at the University of Tennessee. Her research interests include social equity in tourism around the intersectionality of race, gender, sexual orientation and people with disabilities. Additionally, she researches film-induced tourism, implements improvizational theatre games as innovative pedagogy and is a certified qualitative researcher. Lastly, she serves as the Initiative Coordinator and Research Fellow for Tourism RESET – a multi-university and interdisciplinary research and outreach initiative that seeks to identify, study and challenge patterns of social inequity in the tourism industry (www.tourismreset.com).

Catherine Butler is Reader in English Literature at Cardiff University. Her academic books include *Four British Fantasists* (Scarecrow/ChLA, 2006), *Reading History in Children's Books* (with Hallie O'Donovan; Palgrave, 2012) and *Literary Studies Deconstructed* (2018). She has also co-edited several academic collections, including *Modern Children's Literature* (with Kimberley Reynolds, 2014) and essay collections on Roald Dahl and Philip Pullman. She is the author of six novels for children and teenagers, as well as some shorter works. Catherine is Editor-in-Chief of *Children's Literature in Education*.

Christopher Hood is a Reader in Japanese Studies at Cardiff University. His main research interests relate to aspects of the transportation sector in Japan and the way in which airplanes and the shinkansen, in particular, can be used to study features of Japanese society and culture. He is the author of the books *Osutaka: A Chronicle of Loss in the World's Largest Single Plane Crash, Japan: The Basics, Dealing with Disaster in Japan: Responses to the Flight JL123 Crash* and *Shinkansen: From Bullet Train to Symbol of Modern Japan*. He is also the author of the novel *Hijacking Japan*.

Kyungjae Jang is an Associate Professor in the Department of Integrated Global Studies, School of Integrated Arts and Sciences, Hiroshima University. He holds a PhD and an MA in tourism studies from Hokkaido University, and a BA from Korea University. Dr Jang has conducted participatory research on transnational Japanese contents tourism in the

USA, Tunisia, Korea and Taiwan. His previous publications include *Contents Tourism in Japan* (with Philip Seaton, Takayoshi Yamamura and Akiko Sugawa-Shimada) and 'Between soft power and propaganda: The Korean military drama Descendants of the Sun', *Journal of War & Culture Studies* (2019).

Aleksandra Jaworowicz-Zimny received her PhD from Hokkaido University and is currently working at the Cultural Studies Department of Nicolaus Copernicus University, Toruń, Poland. Her research focuses on representation of the Asia-Pacific war-related issues in Japanese fan productions. Recent publications include 'Manga/Anime conventions in Poland: The example of Japanicon 2015' in *International Journal of Contents Tourism* (2017), 'Nazi cosplay in Japan' in *Journal of War & Culture Studies* (2018) and 'Kandō Conservatism – "Moving" war narratives in Japanese online fan videos' in *The Asia-Pacific Journal: Japan Focus* (2018).

Sueun Kim is a Researcher in the Koreanophone Reseach Center, Hankuk University of Foreign Studies. She holds a PhD and an MA in Korean Studies from Hankuk University of Foreign Studies. She is the author of a number of articles on transnational popular culture, including 'Reproduction and Mediation of the Korean wave in Southeast Asia' (in Korean) in the *Southeast Asia Journal*.

Gretchen Larsen is an Associate Professor in Marketing at Durham University. Gretchen's research is located within interpretive and critical consumer research, at the intersection of consumption, markets and the arts. In particular, she seeks to understand how the position of the consumer in a sociocultural world is constructed, performed, interpreted and questioned through the arts. This research examines the importance of arts and sound consumption in helping consumers make sense of their world, and how those involved in producing and distributing arts (marketers, artists, policy makers) can ensure that these benefits are delivered.

Shinobu Myoki is an Associate Professor at the Graduate School of International Cultural Studies in Tohoku University. She received her PhD in Sociology from the University of Tokyo. Specializing in Gender Studies and the Sociology of Leisure, her current research topics are the *shufu ronsō* ('housewife controversy') and erotic museums in the context of women's socioeconomic change in postwar Japan. Myoki is the author of *Confrontation Among Women in Postwar Japan* (2009, in Japanese) and *The Culture of Japan's Erotic Museums (Hihōkan)* (2014, in Japanese). Her most recent work is on the representation of female bodies in medical displays, focusing on Europe and Japan.

Ranny Rastati is a Researcher at the Research Center for Society and Culture–Indonesian Institute of Sciences (PMB-LIPI). She received her bachelor's in Japanese Studies at University of Indonesia and master's in Communication Studies at University of Indonesia. Publications include popular books such as *Ohayou Gozaimasu* (2014) and *Korean Celebrity Song Triplets: Daehan Minguk Manse* (2015), journal articles on hijab cosplay (2015), cyberbullying (2016), Islamic manga (2017), media literacy (2018) and halal tourism (2018). Her research interests include cosplay, pop culture and media studies. Her current research topics are hijab cosplay as preaching Islam, cosplay as contents tourism and halal tourism. She has also managed a non-profit organization for social activities and voluntary service, Chibi Ranran Help Center, since 2013.

Akiko Sugawa-Shimada, PhD, is a Professor in the Graduate School of Urban Innovation at Yokohama National University, Japan. Dr Sugawa-Shimada is the author of a number of books and articles on anime, manga and Cultural Studies, including *Girls and Magic: How Have Girl Heroes Been Accepted?* (2013; won the 2014 Japan Society of Animation Studies Award, in Japanese), chapters in the books *Japanese Animation: East Asian Perspectives* (2013), *Teaching Japanese Popular Culture* (2016), *Shojo Across Media* (2019) and *Women's Manga in Asia and Beyond* (2019), and as co-author, *Contents Tourism in Japan* (2017). Her website is: http://www.akikosugawa.2-d.jp.

Maree Thyne is an Associate Professor in the Department of Marketing at the University of Otago. Maree's research platform draws on multi-disciplinary theories from marketing and tourism and uses knowledge of consumer psychology to provide innovative insights into consumer attitudes and behaviour, often within an arts framework. She has published in a number of Marketing and Tourism journals and she is on the editorial board for *Tourism Management Perspectives*, *Australasian Marketing Journal*, *International Journal of Culture, Tourism and Hospitality Research* and *Tourism and Hospitality Research*.

Acknowledgements

Takayoshi Yamamura and Philip A. Seaton

This book is the concluding publication of a five-year international research project (2014–2019) supported by a JSPS (Japan Society for the Promotion of Science) grant titled 'International Comparative Research on the Spreading and Reception of Culture through Contents Tourism' (Kiban A, grant number 26243007, grant period 2014–2019, project co-leaders Philip Seaton and Takayoshi Yamamura). This research would not have been possible without the generous financial support of JSPS and all the practical support given by the administrative staff at Hokkaido University and Tokyo University of Foreign Studies, who have helped us manage the grant.

The papers for this edited volume were first presented at the International Symposium 'Transnational Contents Tourism in Europe and Asia' held at Tokyo University of Foreign Studies on 7–10 June 2018. We would like to thank all the symposium attendees, both those whose work is included in this book and those who are publishing their research elsewhere. However, we would like to extend particular thanks to Professor Sue Beeton, President of TTRA, whose support, friendship and encouragement have meant so much to us during the contents tourism project ever since we started working together in 2011.

Throughout the project, many people have provided us with valuable information, suggestions and advice. In particular, we would like to express our sincere gratitude to Nelson H.H. Graburn, Professor Emeritus in Sociocultural Anthropology at the University of California, Berkeley, for his warm and long-standing academic support. In particular, it was a great honour to present our research at an international conference that he organized, 'Contents Tourism: Creativity, Fandom, Neo-Destinations', which was held in the Department of Anthropology, University of California, Berkeley, 10–11 March 2017. We would also like to extend our appreciation to Dean MacCannell, Professor Emeritus in Environmental Design, University of California, Davis, for his invaluable insights expressed at the conference. Philip Seaton thanks Dr Karl Ian U. Cheng Chua for his kind invitation to give a keynote lecture at Ateneo de Manila University, the Philippines, in February 2018, at which first drafts of many of the ideas in the conclusions were presented, and also to Dr Helen Macnaughtan and SOAS, University of London, for the invitation to give the W.G. Beasley Memorial Lecture in March 2019 on the research presented in this book.

Our concept of contents tourism has also been inspired by and developed through discussions with colleagues and students at several conferences, institutions and classes, including TTRA Apac (Travel and Tourism Research Association, Asia Pacific Chapter), BAJS (British Association for Japanese Studies), EAJS (European Association for Japanese Studies), AAA (American Anthropological Association), Institute of Japan Studies (Tokyo University of Foreign Studies), MJSP (Modern Japanese Studies Program, Hokkaido University) and IMCTS (Graduate School of International Media, Communication and Tourism Studies, Hokkaido University). We deeply appreciate their warm friendship, academic support and feedback. There are so many other persons and organizations we have to thank. We have acknowledged them in the text where appropriate.

Finally, we thank all the staff at Channel View Publications for their help and support throughout the whole process of reviewing, editing and publishing. We have done our best to ensure an error-free volume, but any remaining errors are the responsibility of the authors and editors.

Note on names: All names are written in the order given name followed by family name in the text, regardless of their nationalities (however, in citations, the order is as in the original). Macrons indicate long vowel sounds in Japanese names and words.

Foreword

Sue Beeton

In 2011 I was invited to speak at Meiji University in Tokyo at a symposium on ecotourism. Little did I know how significant this very short first visit to Japan would be.

At the symposium, Philip Seaton came up to me and spoke briefly about the work he and a colleague were doing on 'contents tourism' and gave me two books on the topic in Japanese, which certainly fed my curiosity as I could really only look at the pictures owing to my total lack of knowledge of the language. Nevertheless, I was fascinated … Around that time I had been thinking a lot about the relationship between my area of film-induced tourism and a broader popular culture tourism connection, of which film is an aspect. Certainly, Japan's love of manga, anime, J-pop and almost obsessive cosplay along with their world-breaking gaming culture seemed to fit with my growing thoughts on a wider relationship with tourism. Furthermore, the obsession the West was developing for such a culture had become clear to me in recent years, reflecting my own interests.

So began my relationship with Japan, pop culture and many of the authors of the chapters in this book as I began to look more closely at the ways that Japanese pop culture has influenced us in the West and our recreational/tourism activities. Was it simply a case of loving all things Japanese, or was there a deeper connection here? To understand this, I needed to further my very limited understanding of Japan and its culture and heritage, which led me on a powerful personal journey, where I have walked on sacred ancient roads, shared prayers with pilgrims at sacred and beautiful sites, followed the paths of famous artists and poets, pondered protected soundscapes and power sites as well as visiting film sets and sites, theme parks and museums and experienced the urban delights of Tokyo with its robot restaurants and themed cafes. I even stayed at a hotel that celebrated the stories of Godzilla!

One of the outcomes of such studies and experiences was that, while Japan is very different from Australia (and the rest of the Western world), pop culture often transcended such differences as we began to see not only Japanese anime on our screens and cosplay events, but also works that were clearly influenced by this, developed by Westerners for our consumption.

One of the founding principles of 'contents tourism' is its narrative component, immediately bringing to mind not only the narratives surrounding film-induced tourism, but also the increasing awareness of our tourism gurus of tourists being interested in 'experiences' as opposed to 'sites'. By their very nature, experiences contain (or create) a narrative. I saw a direct link here between what had been considered a very Japanese concept of *kontentsu tsūrizumu* to a more transnational notion of contents tourism. The work presented in this book by researchers from around the world not only demonstrates this, but also takes us even further on this fascinating journey, which is truly never-ending.

Introduction: Contents Tourism Beyond Anime Tourism

Takayoshi Yamamura

The aims of this book are to apply the concept of 'contents tourism' in a global context and to establish an international and interdisciplinary framework for contents tourism research. Industry and academia in Japan have used the concept of 'contents' since the 1990s, and in our previous book, *Contents Tourism in Japan: Pilgrimages to 'Sacred Sites' of Popular Culture*, we positioned 'contents' as 'the combination of the creative elements' within works of mediatized popular culture (Seaton *et al.*, 2017: 2). Consequently, we defined 'contents tourism' as 'travel behaviour motivated fully or partially by narratives, characters, locations, and other creative elements of popular culture forms including film, television dramas, manga, anime, novels, and computer games' (Seaton *et al.*, 2017: 3). The players, patterns and history of contents tourism in Japan were illustrated via case studies drawn from all of Japan's 47 prefectures. This book builds on that domestic study and analyses transnational and transmedia perspectives on contents tourism with a particular focus on the dynamic process of *contentsization*, namely the development and expansion of a 'narrative world' through both mediatized adaptation and tourism practice.

Contents Tourism Studies in Japan

The term 'contents tourism' (*kontentsu tsūrizumu*) gained official recognition in Japan when it was defined by the Japanese government in 2005 as 'the addition of a "narrative quality" [*monogatarisei*] or "theme" [*tēmasei*] to a region – namely an atmosphere or image particular to the region generated by the contents – and the use of that narrative quality as a tourism resource' (Ministry of Land, Infrastructure and Transport *et al.*, 2005). In the wake of this definition, two main streams of contents tourism research emerged in Japan. The first was a series of studies supporting the 2005 government policy in the wake of its publication.

The second is so-called 'anime tourism' studies, which originated to describe the phenomenon of 'anime pilgrimage' without any relationship to the government policy, and later on was incorporated into contents tourism studies.

The former group of policy-supporting studies shared the government's goal of achieving tourism and community development using contents in local communities in Japan. The definition of contents tourism invariably followed the 2005 policy document. Studies in this group tend to examine contents according to media format within Japan – such as anime, TV dramas, literature and J-pop – and to describe domestic and local case studies. The first policy-supporting study following the 2005 policy document was Toshiyuki Masubuchi's (2010) book, *Monogatari wo tabi suru hitobito: What is Contents Tourism?* It utilized the government's definition of contents tourism and was the first academic attempt both inside and outside Japan to schematize a framework for contents tourism. Focusing on anime, TV dramas, literature, J-pop and other media formats, Masubuchi presented case studies from the viewpoints of geography, urban/tourism development and industrial promotion. Masubuchi also published two sequels (Masubuchi, 2011, 2014), and he has actively attempted to broaden and showcase the field of contents tourism studies via an almost encyclopaedic approach of introducing many cases in different media formats or genres.

Masubuchi was involved in the creation of the Academy of Contents Tourism research group, which was established on 2 October 2011 with Masubuchi as chairperson. The Academy also follows the definition of contents tourism in the 2005 policy document. According to the Academy's charter, its purpose is to establish methodology for utilizing contents in regions, and to undertake research, analysis and discussions in order to construct sustainable tourism development policy, and disseminate this information to the public through the activities of the academy (Academy of Contents Tourism, n.d.). Masubuchi and Academy members published the book *Kontentsu tsūrizumu nyūmon* [Introduction to Contents Tourism] (Masubuchi *et al.*, 2014). Like Masubuchi's earlier books, this book also consists of independent case studies in media such as anime, film, TV drama, literature, local songs, local idols (i.e. celebrities) and idol stage performances. The authors depict the big picture of contents tourism and list studies according to media format. In that they divide up forms of contents tourism according to media format, the book also adheres fundamentally to government policy and Masubuchi's previous discussions.

This approach also resembles much of the English-language literature regarding the study of 'media tourism' (Reijnders, 2011) and 'media pilgrimage' (Norris, 2013) in that researchers analyse tourism phenomena induced by media products focusing on specific media formats. Stijn Reijnders stated that 'some media products lead to media tourism' that has a 'multimedia character' (Reijnders, 2011: 5), but ultimately he also focuses on media formats such as TV dramas, films and literature and discusses how each format creates 'imaginary places' (Reijnders, 2011: 17). In this sense, the process of creating 'imaginary places' has almost the same meaning as contents tourism as defined by the Japanese government, Masubuchi and the Academy of Contents Tourism.

The second group of researchers in Japan concentrate on 'anime tourism' studies that started with analyses of the 'anime pilgrimage' phenomena in Japanese communities. They tend to describe tourism phenomena or the behaviours of actors such as tourists and the local community using ethnographic methodologies. These studies are more likely to suggest a redefinition of the term contents tourism because they focus on tourism phenomena rather than making arguments relating to policy. Most of the case studies focus on *otaku* (nerd) contents in anime, manga, video games and cosplay events.

Anime tourism studies emerged without any specific reference to the 2005 policy document. Anime pilgrimage (a tourism practice in which anime fans visit location models of or sites related to their favourite anime works) and the expansion of exchanges between anime fans and local communities associated with anime pilgrimage has been a growing phenomenon since the mid-2000s (Okamoto, 2013). The pioneering case was the anime *Lucky Star* (2007) and related exchange between fans and locals in Washimiya town, Saitama prefecture. The first paper analysing the phenomena from a community development point of view was my article, 'Study of birth and development of "sacred place for anime fans"' (Yamamura, 2008; see also Yamamura, 2015). This is recognized as one of the earliest studies on anime pilgrimage. However, the 2008 article just used the term 'anime pilgrimage' and did not mention 'contents tourism'. It can be said that anime pilgrimage was not positioned clearly as part of the contents tourism phenomenon in the article.

In 2011, one year after Masubuchi published the first contents tourism book, in my book *Anime, manga de chiiki shinkō* [Regional Revitalization via Anime and Manga] (Yamamura, 2011) I called anime pilgrimage 'anime tourism' and positioned it as a contents tourism practice. Through comparative analysis of several domestic cases, I defined the three major actors in contents tourism – creators, fans and the local community – and argued that the establishment of an interactive and beneficial collaborative system among these three actors is one of the key issues of contents tourism development (Yamamura, 2011). Accordingly, I defined contents tourism as follows: 'tourism in a region or particular place that functions as media, and where tourists feel its contents (narrative quality) through their five senses. Emotive bonds are created between people, or between people and objects, through the shared narrative world' (Yamamura, 2011: 172–173). This collaboration model, based on the establishment of emotive connections among these three major actors and shared respect for the contents/'narrative worlds', has been applied to other anime tourism case studies to assess or analyse their sustainability (e.g. Otani et al., 2018).

Conversely, Takeshi Okamoto (2013) has analysed the *Lucky Star* case from the perspective of the sociology of tourism and the theory of tourist behaviour. Okamoto focused in particular on the creativity of tourists, as evidenced through their development of creations such as derivative works, to propose the concepts of 'tourist-led contents tourism' and '*n*th creation tourism' (Okamoto, 2013). He analysed how 'anime pilgrims' communicate both with each other and non-fan others, and discussed the formation processes of pilgrim communities and the collaborative relationships between pilgrims and local residents. The biggest achievement of his work was his redefinition of the term 'contents', which had only been used in the contents industry until then, in order to apply the term more easily to contents tourism studies. He defined contents as 'information that has been produced and edited in some form and that brings enjoyment when it is consumed' (Okamoto, 2013: 40–41). This definition was based on the results of his fieldwork during which he conducted in-depth ethnographic research on the processes of re-producing and re-editing contents by fans/tourists at the 'sacred sites' of anime. While we defined contents as 'narratives, characters, locations, and other creative elements of popular culture forms' (Seaton et al., 2017: 3) as mentioned above, Okamoto's definition is also useful for our contents tourism approach in this book. Therefore, we can combine these two to redefine contents as *information – such as narratives, characters, locations and other creative elements – that has been produced and edited in popular culture forms and that brings enjoyment when it is consumed.*

Furthermore, based on his definition of contents, in his edited volume *Kontentsu tsūrizumu kenkyū* [Contents Tourism Research], Okamoto redefined the term 'contents tourism' as a broader concept that refers to 'tourism/travel behaviour motivated by contents, or tourism/community development utilizing contents' (Okamoto, 2015: 10). The book is a collection of papers by researchers from a variety of academic fields based on this broadened concept of contents tourism. As he pointed out in his book, this broader concept is useful for activating 'unrestricted discussions on relationships between contents and tourism', and the book indeed broadened the potential field of contents tourism studies. However, most of the case studies included in the book are anime-related domestic Japanese cases, and the discussions still only focus on Japanese domestic issues, framed by the media formats of the contents.

From the two major streams of contents tourism studies in Japan described, it is clear that both of these streams still mainly focus on domestic Japanese and media-format-based cases. In Japan, the focal points of discussion regarding contents tourism are issues such as policy support, pilgrimage/tourist behaviour and collaboration models among actors. Case studies focusing on specific media and genres are increasing.

However, both streams suffer from the common issue that their discussions are little more than domestic case studies and they tend to emphasize the uniqueness of Japanese culture. In other words, they lack comparative discussions regarding similar non-Japanese cases and do not attempt to establish a research framework from an international point of view. This is a severe limitation of contents tourism research in Japan. This is possibly because most of these studies have been undertaken by individual Japanese researchers within their own fields and without an international research framework. Whatever the reason, existing contents tourism research in Japan has not satisfactorily discussed the transmedia and transnational phenomena of contents, which are used across media formats, instantly consumed across borders and induce transnational tourism experiences. This means that, in Japan, there remains an incomplete understanding of the dynamic transmedia and transnational aspects of contents tourism.

Re-framing Contents Tourism Approaches from an International Perspective

The concept of 'contents' emerged in Japan. Yet as discussed, Japanese scholars tend to treat contents tourism in Japan (as well as Japanese pop culture) as a special case separated from related cases in other countries. However, the media mix phenomenon and transnational consumption of contents are now evident not only in Japan, but also worldwide. Therefore, it is necessary to construct an approach and framework for contents tourism that takes into account its transnational and transmedia perspectives using the achievements and limitations of previous contents tourism studies in Japan.

Beeton *et al.* (2013) published a paper in English that introduced the concept of contents tourism in the international context to an academic audience outside of Japan. This book chapter was the starting point of our international contents tourism research project and laid the groundwork for this book. Beeton *et al.* map a transnational perspective of contents tourism studies, which previous studies in Japan had not yet done, and discussed cases in Japan, South Korea, Hong Kong, Bollywood and Australia. We highlighted the difference between film-induced tourism and contents tourism, noting that the latter 'cuts across particular media forms to focus on the narratives: *monogatarisei* (narrative quality)' (Beeton *et al.*, 2013: 146). Moreover, we presented two important processes in contents tourism, namely the mediatization of 'sites' and mediatization of the tourist 'experience', placing the media mix and multi-use of contents at the heart of contents tourism (Beeton *et al.*, 2013: 150). We concluded that physical sites can be considered as a form of media within the contents tourism approach. As such, the article laid the foundations for transnational contents tourism research.

Thereafter, Beeton has continued to position contents tourism within the framework of film-induced tourism (Beeton, 2015, 2016). In particular, in 2016, she added a section about contents tourism in the second edition of her seminal book *Film-Induced Tourism* and integrated the concept fully within existing English-language discourse (Beeton, 2016: 31–33). Following Beeton's book, in our 2017 book *Contents Tourism in Japan* (Seaton et al., 2017) we situated contents tourism firmly within the international literature on tourism, media and media-induced tourism and provided a historical overview of tourism practices meeting the definition of 'travel behaviour motivated fully or partially' by mediatized works of popular culture. This redefinition of contents tourism removed the theory from its previously Japan-specific context and enables non-Japanese cases to be examined using the contents tourism framework. Furthermore, a more specific explanation of what may be considered contents was given. 'Contents' fall within 'popular culture', but with three main features: (a) 'popular culture is mediatized' and 'does not include non-mediatized culture such as cuisine'; (b) 'the mediatized culture contains creative elements and is produced primarily as entertainment'; and (c) 'by being "of the people," popular culture is not the work of government and there are no distinctions between "art" and "pop culture"' or '"high" and "low"' (Seaton et al., 2017: 4–5). By doing so, Seaton et al. shift the attention from media formats to 'the contents' and argue that the approach is better suited than existing format-based approaches to analysing the tourist behaviours of fans who visit a narrative world in the mixed-media age because the use of contents beyond their original media formats is now widespread (Seaton et al., 2017: 2, 265).

Subsequently, other scholars have begun utilizing the concept of contents tourism in their work. Tzanelli (2019) adopted the notion of contents tourism as a replacement for the term 'cinematic tourism' and described it as 'a form of pilgrimage to foreign and home lands, which is both directed towards heritage and creates forms of heritage anew' and 'an ancient and yet modern form of mobility standing at an intersection of emotion and reason, to which business design adds little in terms of experience' (Tzanelli, 2019: 3–4). She also stressed the advantage of the contents tourism approach as 'contents tourism allows us to trace the ways contemporary tourist industries and localities create the atmosphere of the tourist destination' (Tzanelli, 2019: 4). In her book, atmospheres refer to 'the ways place is constructed in multi-sensory ways and experienced by those who inhabit/visit it, so it has both natural, material, phenomenal and cultural dimensions' (Tzanelli, 2019: 4). Her work expands the horizons of contents tourism research by reconstructing the previous framework from a new perspective based on cultural sociology.

Within contents tourism other key concepts are content multi-use and the media mix, on which topics there have been several important pioneering studies such as those by Jenkins (2006), Steinberg (2012),

Condry (2013) and Galbraith and Karlin (2016). Jenkins focused on 'the flow of content across multiple media platforms, the cooperation between multiple media industries, and the migratory behaviour of media audiences who will go almost anywhere in search of the kinds of entertainment experiences they want' and named these phenomena 'convergence' (Jenkins, 2006: 2). Moreover, he introduced the concept of 'transmedia storytelling' as a specific example of the convergence and explained that:

> a transmedia story unfolds across multiple media platforms, with each new text making a distinctive and valuable contribution to the whole. In the ideal form of transmedia storytelling, each medium does what it does best – so that a story might be introduced in a film, expanded through television, novels, and comics. (Jenkins, 2006: 95–96)

The studies by Steinberg, Condry, and Galbraith and Karlin are significantly influenced by Jenkins's work, especially regarding the idea of transmedia storytelling. Steinberg (2012) developed media mix theory by analysing mainly the anime industry in Japan from the perspectives of industrial strategy and media history. In contrast, Condry (2013) discussed the media mix including the practices of fans relating to Japanese anime characters which are disseminated across various media. However, both studies positioned worlds or original works as occupying the most significant position and placed the multitude of other media platforms in parallel under them as subordinate. The edited volume by Galbraith and Karlin (2016) shifts the view of convergence away from anime and manga and addresses issues such as media in the wake of the Great East Japan Earthquake of 2011, and convergence relating to Korean dramas and celebrity culture.

In short, the literature on the media mix has addressed how creators utilize media platforms and how consumers obtain 'entertainment experiences' 'across multiple media platforms' (Jenkins, 2006: 2). In this book, we are developing these ideas further, but we will focus on the contents themselves rather than the media platforms. In particular, we aim to clarify the dynamism in the process of narrative world development and expansion, which is achieved thorough the re-editing/re-creating of contents across various media formats through analysis of tourism practice and experience.

'Contentsization': A Dynamic Process of Narrative World Development/Expansion through both Mediatized Adaptation and Tourism Practice

As mentioned above, the advantage of the contents tourism approach is that it places the focus not on media formats but on contents that are multi-used across media formats and generate a series of tourist behaviours/experiences motivated by the contents. In this sense, the processes of contents multi-use are the core issues to be discussed in contents tourism studies. For example, after a piece of literature is created as an original work by a writer, the literature may be re-created into a TV drama and then a film by other creators. Moreover, fans create works of fan fiction and re-enact scenes from screen adaptations during cosplay. In these ways, the original works are re-interpreted, re-edited and re-performed in other works or forms of expression both on-site and off-site, both by professional creators and amateurs/fans. The original contents, therefore, stimulate renewed content production and the contents or narrative worlds expand as a result. The more the contents are re-edited/re-created, the more the narrative worlds expand, and the resultant complex set of contents induces tourism behaviours in multiple ways. Contents tourism is characterized by these complex and expansive aspects.

There are many examples of such dynamism in the process of narrative world expansion. For example, Seaton *et al.* examined *The Tale of Genji*, the oldest extant work of literature in Japan, to illustrate how narrative worlds have been re-interpreted and re-edited in multiple media formats such as novels, Noh theatre and paintings since as early as the Muromachi period (1336–1573), and these expanded narrative worlds still induce tourism today (Seaton *et al.*, 2017: 25, 78–80). Yamamura (2011) also used the history of Yukimura Sanada, a military commander in the 16th century, to show that, since the novelization of his life in the Edo period (1603–1868), related narrative worlds have been re-created in a variety of media formats such as TV dramas, video games and anime. As a result of the multi-use and recreation of the contents, the narrative worlds that belong to the local community in Ueda city, Nagano prefecture, have expanded and the attractiveness of related sites as contents tourism destinations has also been increasing (Yamamura, 2011: 193–194).

Such issues have often been discussed in the context of the perceived uniqueness of Japanese culture by scholars focusing on Japan. However, in this book, we demonstrate that similar cases can be seen not only in Japan but also all over the world. For instance, *Romance of the Three Kingdoms*, a novelized Chinese history of the second and third centuries, is another example. Since its novelization in the 14th century (Ming dynasty era), the narrative world has been 'contentsized' repeatedly over the centuries, not only in China but also worldwide, by many creators in forms such as literature, puppet theatre, video games, TV dramas, anime and cinema. With this contentsization, the evolving narrative world continues to expand the tourist imagination of China and attracts tourists to the country. Outside Asia, the Sherlock Holmes books by Sir Arthur Conan Doyle provide another good example of this phenomenon.

Such dynamism in the process of narrative world expansion has been discussed mainly in the field of media studies since the 2000s using the concepts of 'adaptations' (Murray & Weedon, 2011), 'transmedia storytelling' (Jenkins, 2003, 2006), 'transtextual worlds' (Hills, 2015), 'spreadable media' (Jenkins *et al.*, 2013) and 'media mix' (Otsuka, 2014; Steinberg, 2012, 2015). These concepts have been used to describe the dynamism in the processes by which contents are adapted, remade and consumed beyond the media format of the original work. The contents reconstruction process discussed in studies of contents tourism are congruent with these concepts in media studies.

However, there is an important difference that distinguishes narrative world expansion in contents tourism and the other related concepts from media studies just identified. As already mentioned above, the contents tourism approach considers physical sites to be 'media', and therefore regards tourism practices as a driving force for the transmedia expansion of contents and narrative worlds. In contrast, the concepts from media studies – such as adaptations, transmedia storytelling, transtextual worlds, spreadable media and media mix – do not do so. Furthermore, there are very few examples of research within tourism studies that focus on physical sites as media. Consequently, we will look at the relationship between the multi-use of contents and tourist phenomena/practices by considering tourist sites as media and viewing tourism practices as a driving force for the transmedia expansion of contents. This is the distinctive contribution of a contents tourism approach.

In these ways, we clarify the processes of the development and expansion of narrative worlds, which we call 'contentsization'. We first proposed the term contentsization in a special edition of the *Journal of War & Culture Studies* (Seaton, 2019: 5; Yamamura, 2019: 11) to refer to the dynamic aspects of expansion of contents or 'narrative worlds' through the process of re-editing and converting a creator's work for simultaneous use in a variety of media formats, such as manga, games, live-action films, anime and so on. In the light of the above discussion, and in order to advance the transnational and transmedia discussions of contents tourism, we can further develop the notion of contentsization with particular emphasis on physical sites as media, and by regarding tourism practices as a driving force for transmedia expansion of contents. Consequently, we are presenting an alternative, refined definition of contents tourism that builds on the definition in Seaton *et al.* (2017: 3) cited at the beginning of this Introduction, and contentsization as used by Seaton (2019: 5) and Yamamura (2019: 11). The modified definition is as follows:

> Contents tourism is a dynamic series of tourism practices/experiences motivated by contents (defined above as 'information – such as narratives, characters, locations, and other creative elements – that has been produced and edited in popular culture forms and that brings enjoyment when it is consumed'). Contents tourists access and embody 'narrative worlds' that are evolving through 'contentsization', namely the continual process of the development and expansion of the 'narrative world' through both mediatized adaptation and tourism practice.

In other words, the distinguishing characteristic of contents tourism and contentsization is the focus on the role that tourists play in convergence culture via their travel behaviours at real-world sites (which, as described above, are treated as media) related to the narrative world.

This book further develops contents tourism theory to provide a versatile international and interdisciplinary framework not only based on the above-mentioned series of previous studies but also in accordance with the conditions of transnational expansion of narrative worlds and cultural exchange in an age of advancing informatization and mobility. More specifically, we will focus on the relationship between transnational and transmedia processes of expansion (via creation, re-creation, adaptation and consumption) of narrative worlds and a series of behaviours in which tourists attempt to embody (or emotively/bodily access) narrative worlds through on-site tourist experiences. In other words, it can be said that tourism practices and experiences can expand narrative worlds. Moreover, considering such behaviours as basic needs of narrative world consumption common to all humankind, all cases, regardless of their place or country of origin, can be treated on an equal basis.

Applying the Contents Tourism Approach to Multiple Dimensions of Contemporary Tourism Practices

In the 'Information Age', there is an exponential increase in the volumes of information available, and increased mobility of people, information and contents across borders. Tourism practices are also becoming more mediatized. Tourists are trying not just to visit and see places themselves, but to consume, embody, reinterpret and reconstruct their favourite narrative worlds through on-site tourist experiences. In other words, many tourism phenomena are acquiring the characteristics of contents tourism. Therefore, using the above-mentioned alternative contents tourism definition and approach focusing on the 'contentsization' process, an increasing number of tourism phenomena can be analysed, including those that have not been studied before from a contents tourism perspective.

One of the typical examples is the cosplay that Ranny Rastati discusses in Chapter 9. Cosplay is generally defined as a creative practice or performative activity 'in which fans produce their own costumes inspired by fictional characters' and 'a form of appropriation that transforms and actualizes an existing story in close connection to the fan community and the fan's own identity' (Lamerichs, 2011: Chapter 0.1). It has been studied mainly in the fields of fashion and gender studies (Narumi, 2009), media studies focusing on fan culture (Jenkins, 2006; Lamerichs, 2011), and cultural studies related to the transmission and acceptance of pop culture (Iwabuchi, 2001). Recently, along with the activation of studies on the boundaries between media experience and tourism practice, scholars working in tourism studies also started to address cosplay from around 2010 in the context of contents tourism (Beeton *et al.*, 2013; Kikuchi & Shizuka, 2017). However, there are still very few studies on this topic and they have not defined cosplay clearly as contents tourism yet.

By applying the alternative contents tourism definition and approach, we can clearly include cosplay within contents tourism practices for the following two reasons. First, historically speaking, cosplay itself has developed through the transnational and transmedia process of cultural diffusion and absorption. Second, we can see the dynamic process of consumption, re-interpretation and recreation of contents by cosplayers in their cosplay practices.

Regarding the first point, the history of cosplay 'actually blends the Japanese and North American contributions' (Winge, 2006: 66) and we can see the transnational and transmedia process of cultural diffusion and absorption clearly there, as Bruno (2002) pointed out:

> Twenty years ago, the first fan costumers were seen in Japan at a small comic expo known as Comiket or Comic Market. They were simply wearing t-shirts on which they'd drawn their favorite characters. The following year, 1983, the first actual costume was worn by someone dressed at [sic] Lum from Urusei Yatsura which was airing in Japan at the time. In 1984, Mr Takahashi was sent to Worldcon in Los Angeles to cover the events for various magazines back in Japan. [...] he was amazed by what he saw. Many people dressed as their favorite characters from Star Trek, Star Wars and even in their own costumed creations. He was particularly impressed by the Masquerade. [...] When he returned to Japan, he wanted to impress upon his readers the magnificence of what he had seen in hopes [sic] that that they could adapt American costuming practices into their own culture. [...] He also wanted something that was neither Japanese nor American, but a combination of both to show the blending of the American costuming tradition with Japanese culture. He finally settled on 'Cosplay' by using the Japanese habit of shortening words into easier to say bits on 'Costume Play'. Thus was born Cosplay. (Bruno, 2002)

Regarding the second point, as Rastati discusses in detail in Chapter 9, we can clearly see the dynamic process of consumption, re-interpretation and recreation of contents by cosplayers as prosumers of contents in their cosplay practices, such as creating costumes, searching for locations for cosplay, posing, taking photographs, organizing events, transmitting their photos through SNS and cross-referencing themselves as fictional cosplayers in fictional narrative worlds.

In this manner, applying the contents tourism approach focusing on 'contentsization' to cosplay practice, we can illuminate a dynamic process of 'contentsization' within it and analyse the relationship between the transnational and transmedia processes of the development and expansion of narrative worlds. That is to say, cosplay is one of the on-site performative practices contents tourists use to enact and reconstruct their favourite narrative worlds in the real world by focusing on characters. Similarly, in this book, we will analyse multiple dimensions of tourism practices motivated by contents, which are diversifying rapidly in the era of informatization and mobility, from a contents tourism perspective.

Structure of the Book

In order to apply the concept of contents tourism in a global context and to establish an international and interdisciplinary framework for contents tourism research, we analyse transnational and transmedia perspectives on contents tourism using three major approaches to understand the dynamic process of *contentsization*. The first is to focus on literary worlds as the core of the contents and trace the process of multi-use/multi-consumption of them related to specific physical sites. The second is to centre on the ways in which physical sites acquire 'sacredness', or the fan rituals that are formed through the performance of fans or locals, in order to understand the processes of the development and expansion of the narrative worlds. The third is to place emphasis on personal journeys and their practices/experiences to see the connection/linkage between specific destinations and the minds/imaginaries of fans. Based on these three main approaches, this book consists of three parts: Part 1 (Chapters 1–4), The Contentsization of Literary Worlds; Part 2 (Chapters 5–9), Tourist Behaviours at 'Sacred Sites' of Contents Tourism; and Part 3 (Chapters 10–13), Contents Tourism as Pilgrimage.

Part 1 (Chapters 1–4) introduces and discusses contentsization of literary worlds. Many sustainable cases of contents tourism have the written word, whether historical tales, recorded history or literature, as the works that established the original narrative world. Whether the original motif of a narrative world is based on historical facts or purely the creator's imagination, we hypothesize that depicting the narrative world from the very beginning using the written word is one of the key factors affecting the subsequent contentsization of the narrative world across a variety of media formats. This can be attributed to the fact that the written word is an archaic medium. The human experience of consumption of the written word is less affected by the march of technological advance, which particularly affects audio-visual formats like cinema, so being based on the written word gives more of a timeless quality to the narrative world. The written word acts as an anchor stabilizing the fundamental characteristics of the narrative world, which may then be contentsized by subsequent generations according to the technological means at their disposal. In other words, textual information has strong power to produce ideas in the minds of subsequent generations of creators. Consequently, narrative worlds emanating from the written word are at the heart of many contentsization processes.

In this section, the first two chapters focus on novelists as starting points to see the contentsization processes of their created narrative worlds, and the next two chapters focus on the contents themselves to analyse cross-referencing processes among media. Philip Seaton (Chapter 1) and Maree Thyne and Gretchen Larsen (Chapter 2) discuss the English novelists Jane Austen and the Brontës, respectively. Then Aleksandra Jaworowicz-Zimny (Chapter 3) and Takayoshi Yamamura (Chapter 4) focus on the specific versions of contents (respectively the *Witcher* novels and games, and the novel and anime *Heidi*). These chapters suggest the importance of the written word itself and physical experiences at sites based on the written word in the sustainable process of contentsization, even if the media formats and digital media experiences are highly developed.

Part 2 comprises five chapters (Chapters 5–9) that focus on the ways in which physical sites acquire 'sacredness', and how fans' rituals are formed through the performances of fans or locals. Performativity or performance is one of the key issues that has been widely discussed in the field of tourism studies since around 2010 (for example, 'performative authenticity' by Knudsen & Waade, 2010; the 'performance turn' by Larsen, 2010) and performance can be seen in a wide variety of ways. Therefore, in this part, the authors analyse various tourist or local performances in the context of contents tourism. The first two chapters (Chapters 5 and 6) analyse the ways in which specific sites acquire 'sacredness' thorough tourist or local performances. In Chapter 5, Catherine Butler introduces the case of the Cotswolds in England, focusing on the tourist gaze; and in Chapter 6 Shinobu Myoki analyses small towns in Japan and Taiwan (Sakaiminato, Yamashiro, Tono in Japan and Xitou in Taiwan), focusing on interpretation and communication using *yōkai* (mythical creatures or monsters) contents. The last three chapters analyse the ways in which fan rituals are formed through a variety of contents tourism practices. Kyungjae Jang (Chapter 7) discusses the religious imagination in the contents tourism rituals of Korean fans of Japanese contents; Akiko Sugawa-Shimada (Chapter 8) presents fans as active consumers and producers through the emerging phenomena of 2.5-D theatre (theatrical adaptations of manga, anime and games) in Japan; and Ranny Rastati (Chapter 9) analyses the acceptance of cosplay culture and practices of cosplayers in Indonesia to discuss the potential of cosplay as contents tourism. Through these five chapters, the dynamic and various processes of reinterpreting and expanding the narrative world are shown.

Part 3 has four chapters that focus more on the personal level of tourist experiences as a journey or pilgrimage to clarify the connection/linkage between specific destinations and the minds/imaginaries of fans. Chapters 10 and 11 deal with more objective analysis of the tourist experience and imaginaries. Sueun Kim (Chapter 10) discusses destination images of Korea for Thai tourists motivated by Thai films using the concept of 'Koreanness'; and Christopher Hood (Chapter 11) discusses contents tourism practices in 'plane sight' to introduce a new concept of 'wrapping' planes as a form of 'industrial cosplay'. Then, Chapters 12 and 13 describe personal experiences as contents tourists through autoethnography. Stefanie Benjamin (Chapter 12) describes in detail her travel to New Mexico as a female pilgrim of pop culture and Sue Beeton (Chapter 13) writes of her deep travel experience following the contents related to two poets, A.B. 'Banjo' Paterson in Australia and Basho Matsuo in Japan.

Through these three parts analysing a wide range of aspects of transnational contents tourism, we have tried to present an international and interdisciplinary framework for contents tourism research applying the concept of 'contents tourism' in a global context. The book ends with a set of conclusions by Philip Seaton which draw together the key themes and propose a set of policy implications for achieving 'successful' and sustainable contents tourism. We believe that the concept of contents tourism is becoming ever more useful for understanding the complicated mediatized aspects of tourism phenomena in the 21st century, when physical spaces and cyber space are connected and integrated ever more closely. By concluding with a short section on the *Ghost in the Shell* and Harry Potter phenomena, this book constitutes both a culmination of one process, namely of shifting contents tourism research from its Japanese origins to the international and transnational stage, and the beginning of a new stage: the potential for using the contents tourism approach to gain new insights into tourism phenomena related to iconic sets of global contents.

References

Academy of Contents Tourism (n.d.) Setsuritsu shui. See http://contentstourism.com/seturitushi.html (accessed January 2019).
Beeton, S. (2015) *Travel, Tourism and the Moving Image*. Bristol: Channel View Publications.
Beeton, S. (2016) *Film-Induced Tourism* (2nd edn). Bristol: Channel View Publications.
Beeton, S., Yamamura, T. and Seaton, P. (2013) The mediatisation of culture: Japanese contents tourism and pop culture. In J. Lester and C. Scarles (eds) *Mediating the Tourist Experience: From Brochures to Virtual Encounters* (pp. 139–154). Farnham: Ashgate.
Bruno, M. (2002) Cosplay: The illegitimate child of SF masquerades. *Glitz and Glitter Newsletter, Millennium Costume Guild*, October. See http://millenniumcg.tripod.com/glitzglitter/1002articles.html (accessed January 2019).
Condry, I. (2013) *The Soul of Anime*. Durham, NC: Duke University Press.
Galbraith, P.W. and Karlin, J.G. (2016) *Media Convergence in Japan*. Ebook: Kinema Club.
Hills, M. (2015) From 'Multiverse' to 'Abramsverse': *Blade Runner, Star Trek*, multiplicity, and the authorizing of Cult/SF worlds. In J.P. Telotte and G. Duchovnay (eds) *Science Fiction Double Feature: The Science Fiction Film as Cult Text* (pp. 21–37). Liverpool: Liverpool University Press.
Iwabuchi, K. (2001) *Transnational Japan*. Tokyo: Iwanami Shoten.
Jenkins, H. (2003) Transmedia storytelling: Moving characters from books to films to video games can make them stronger and more compelling. *MIT Technology Review*, 15 January. See https://www.technologyreview.com/s/401760/transmedia-storytelling/ (accessed May 2018).

Jenkins, H. (2006) *Convergence Culture: Where Old and New Media Collide*. New York: New York University Press.
Jenkins, H., Ford, S. and Green, J. (2013) *Spreadable Media: Creating Value and Meaning in a Networked Culture*. New York: New York University Press.
Kikuchi, E. and Shizuka, M. (2017) 'Machi-Cos' as a contents tourism phenomenon: From a case study of 'We Love Cosplay in Miyashiro 2016'. *Contents Tourism Review* 4 (March 2017), 24–34.
Knudsen, B.T. and Waade, A.M. (2010) Performative authenticity and spatial experience: Rethinking the relation between travel, place and emotion. In B.T. Knudsen and A.M. Waade (eds) *Re-investing Authenticity: Tourism, Place and Emotions* (pp. 1–19). Bristol: Channel View Publications.
Lamerichs, N. (2011) Stranger than fiction: Fan identity in cosplay. *Transformative Works and Cultures* no. 7; DOI: https://doi.org/10.3983/twc.2011.0246.
Larsen, J. (2010) Goffman and the tourist gaze: A performative perspective on tourism mobilities. In M.H. Jacobsen (ed.) *The Contemporary Goffman* (pp. 313–332). New York: Routledge.
Masubuchi, T. (2010) *Monogatari wo Tabi Suru Hitobito. What is Contents Tourism?* Tokyo: Sairyusha.
Masubuchi, T. (2011) *Monogatari wo Tabi Suru Hitobito 2. Gotōchi Song no Arukikata*. Tokyo: Sairyusha.
Masubuchi, T. (2014) *Monogatari wo Tabi Suru Hitobito 3. Kontentsu Tsūrizumu to shite no Bungaku Meguri*. Tokyo: Sairyusha.
Masubuchi, T., Mizoo, Y., Yasuda, N., Nakamura, T., Hashimoto, H., Iwasaki, T., Yoshiguchi, K. and Asada, M. (2014) *Kontentsu Tsūrizumu Nyūmon*. Tokyo: Kokon Shoin.
Ministry of Land, Infrastructure and Transport, Ministry of Economy, Trade, and Industry and Agency for Cultural Affairs. (2005) *Eizō tō Kontentsu no Seisaku, Katsuyō ni Yoru Chiiki Shinkō no Arikata ni Kansuru Chōsa*. See http://www.mlit.go.jp/kokudokeikaku/souhatu/h16seika/12eizou/12_3.pdf (accessed June 2018).
Murray, S. and Weedon, A. (2011) Beyond medium specificity: Adaptations, cross-media practices and branded entertainments. *Convergence: The International Journal of Research into New Media Technologies* 17 (1), 3–5; DOI: https://journals.sagepub.com/doi/pdf/10.1177/1354856510386861.
Narumi, H. (ed.) (2009) *Kosupureka Suru Shakai: Sabukaruchā no Shintai Bunka*. Tokyo: Serica Shobō.
Norris, C. (2013) A Japanese media pilgrimage to a Tasmanian bakery. *Transformative Works and Cultures* 14. See http://journal.transformativeworks.org/index.php/twc/article/view/470/403 (accessed May 2018).
Okamoto, T. (2013) *n-th Creation Tourism: Anime Seichi Junrei/Kontentsu Tsūrizumu/Kankō Shakaigaku no Kanōsei*. Ebetsu: Hokkaido Bokengeijutsu Shuppan.
Okamoto, T. (ed.) (2015) *Kontentsu Tsūrizumu Kenkyū: Jōhō Shakai no Kankō Kōdō to Chiiki Shinkō*. Tokyo: Fukumura Shuppan.
Otani, N., Matsumoto, A. and Yamamura, T. (2018) *Kontentsu ga Hiraku Chiiki no Kanōsei*. Tokyo: Dobunkan.
Otsuka, E. (2014) *Media Mix-ka Suru Nihon*. Tokyo: East Press.
Reijnders, S. (2011) *Places of the Imagination: Media, Tourism, Culture*. Farnham: Ashgate.
Seaton, P. (2019) Introduction: War, popular culture, and contents tourism in East Asia. *Journal of War & Culture Studies* 12 (1), 1–7.
Seaton, P., Yamamura, T., Sugawa-Shimada, A. and Jang, K. (2017) *Contents Tourism in Japan: Pilgrimages to 'Sacred Sites' of Popular Culture*. New York: Cambria Press.
Steinberg, M. (2012) *Anime's Media Mix: Franchising Toys and Characters in Japan*. Minneapolis, MN: University of Minnesota Press.

Steinberg, M. (2015) *Naze Nihon wa <Media Mix Suru Kuni> Nanoka*. Tokyo: Kadokawa.
Tzanelli, R. (2019) *Cinematic Tourist Mobilities and the Plight of Development: On Atmospheres, Affects, and Environments*. Abingdon: Routledge.
Winge, T. (2006) Costuming the imagination: Origins of anime and manga cosplay. *Mechademia* 1 (pp. 65–76). Minneapolis, MN: University of Minnesota Press.
Yamamura, T. (2008) Study of birth and development of 'sacred place for anime fans': Discussion of tourist promotions based on animated work. *The Journal of International Media, Communication, and Tourism Studies* 7, 145–164.
Yamamura, T. (2011) *Anime, Manga de Chiiki Shinkō: Machi no Fan wo Umu Kontentsu Tsūrizumu Kaihatsuhō*. Tokyo: Tokyo Horei Shuppan.
Yamamura, T. (2015) Contents tourism and local community response: Lucky Star and collaborative anime-induced tourism in Washimiya. *Japan Forum* 27 (1), 59–81.
Yamamura, T. (2019) Cooperation between anime producers and the Japan Self-Defense Force: Creating fantasy and/or propaganda? *Journal of War & Culture Studies* 12 (1), 8–23.

Part 1
The Contentsization of Literary Worlds

1 The Contents Tourism of Jane Austen's American Fans

Philip Seaton

The UK is one of the leading nations for literary tourism. Many famous writers from William Shakespeare to J.K. Rowling have penned works that have induced major tourism phenomena. However, in the digital age, the written word frequently reaches consumers simultaneously via multiple other formats – such as screen adaptations, graphic novels and theatre – and fans themselves are engaged as prosumers – consumers who also produce spin-off works, events, fan websites and travel books/blogs. When such multi-use occurs, an author (Jane Austen), character (Harry Potter) or franchise (*Game of Thrones*) can become a byword for a broader 'narrative world'. When this narrative world induces tourism, it becomes more appropriate to talk of a 'tourism imaginary' (Chronis, 2012), 'places of the imagination' (Reijnders, 2011) or contents tourism (Beeton *et al.*, 2013; Seaton *et al.*, 2017), rather than 'literary tourism'.

While accepting that examples of discrete media format tourism do exist – *Lord of the Rings* tourism in New Zealand is film-induced tourism, and travel to Thomas Hardy sites by someone who has only ever read the books is literary tourism – such examples are getting ever harder to identify in our increasingly media-saturated world. In the Introduction to this book, Takayoshi Yamamura identified the concept of 'contentsization' as a 'continual process of the development and expansion of the "narrative world" through both mediatized adaptation and tourism practice'. This chapter argues that the current popularity of Jane Austen, one of English literature's most revered authors, is more than a phenomenon of 'convergence' (Jenkins, 2006) or 'transtextual worlds' (Hills, 2015), whereby the original novels have triggered multiple adaptations in various formats by both professional media producers and amateur fans. Tourism and the fan performances at tourist events (such dressing up in period costumes at the Jane Austen Festival in Bath) play a central role in the process of contents expansion and the ongoing development of 'Austen's world'. In this sense, the current Austen boom is an archetypal example of contentsization.

So, while Austen-related tourism has typically been treated as literary tourism (for example Crang, 2003; Herbert, 2001; Wells, 2011: Chapter 4) – and I would concur that in its early stages it was primarily 'literary tourism' – there has been a paradigmatic shift in the nature of Austen-related tourism towards contents tourism, with the year 1995 as the key moment in that transition.

The chapter starts with an historical overview of this evolution in 'Jane's Fame' (Harman, 2009) and its relationship to Austen-related tourism. Then, with a focus on Austen's American fans, the motivations for Austen-related contents tourism are examined, and in particular the ways in which Austen-related tourism fits into the broader imaginations and itineraries of American visitors to Britain.

Visiting Jane's World: From Literary Tourism to Contents Tourism

When Jane Austen died in 1817, she was relatively unknown. She had some ardent admirers in the mid-19th century, but 'Janeitism' did not emerge until the 1880s, after J.E. Austen-Leigh's biography *A Memoir of Jane Austen* and the first collected edition of her works were published in 1870 and 1882, respectively (Johnson, 2011: 232). Her works were the subject of debate within literary circles in the early 20th century, but the popularized 'Austenmania' (Todd, 2015: Chapter 9) or 'Jane Austen™' (Harman, 2009: Chapter 7) of today largely dates from the 1990s. The cause of this modern wave of popularization is screen adaptations. The first Austen film was the 1940 *Pride and Prejudice* starring Lawrence Olivier and Greer Garson. Thereafter, there were radio plays and television serials that were theatrical in nature and adhered closely to Austen's original dialogue. However, in 1995, there were three major screen adaptations: the BBC's *Pride and Prejudice*, Columbia's *Sense and Sensibility* and a BBC telefilm of *Persuasion* (Sutherland, 2011: 218–219). Then, with two adaptations of *Emma* appearing in 1996, four of Austen's six completed novels had been adapted for large and small screen and released within two years (Higson, 2011: 132–133). In particular, *Pride and Prejudice* – with its iconic scene of Elizabeth Bennet encountering a wet-shirted Darcy (Colin Firth) after a swim in the lake in the grounds of Pemberley – ignited Austenmania. The thesis of this chapter is that, mirroring the development of Austenmania as a whole, tourism relating to Austen underwent a fundamental transformation in 1995. Up until 1995, Austen-related travel was primarily literary tourism. However, after 1995, travel to a tourism imaginary we may call 'Austen's world' has been inspired by screen adaptations and derivative works as well as the novels in a phenomenon better termed contents tourism.

Austen-related tourism dates back to the 19th century. Literary tourism was a recognizable phenomenon during Austen's lifetime and by a few decades after her death, Austen's fans were already travelling to places connected to her life or depicted in her novels. 'Appreciative readers in the

1850s began to search for their favourite author's burial place in Winchester Cathedral, "the shrine of Jane Austen", one woman called it', writes Hazel Jones, who also describes a journey to sites where Austen lived undertaken by Constance and Ellen Hill in 1901 (Jones, 2014: 157). Readers also engaged in location hunting, for example, Alfred Lord Tennyson made a trip in 1867 to find the location in Lyme Regis made famous by a scene in *Persuasion*, in which Louisa Musgrove fell down some steps (Jones, 2014: 159). These trips were all taken long before any adaptations or the commercial touristification of Austen-related sites.

The touristification of Jane Austen began in 1949 with the opening of the Jane Austen's House Museum in Chawton. Austen lived in this house from 1809 to just before her death in 1817, and it was where she wrote or revised for publication her major works. This is a key site of pilgrimage for Austen fans, along with her birthplace in Steventon (Figure 1.4) and her grave in Winchester Cathedral (Figure 1.1). The museum operates as a site of artefact preservation and scholarship as much as a tourist site, along with other sites holding collections of Austen manuscripts, letters and memorabilia, such as the Bodleian Library in Oxford and the Morgan Library in New York. Research by David Herbert in 1993 and 1994 indicated a strong element of literary tourism at Chawton: in a sample of 223 respondents, 60% of visitors interviewed had read three or more of Austen's novels (Herbert, 2001: 322).

Figure 1.1 Jane Austen's grave in Winchester Cathedral (September 2017). In the alcove at the back are messages to Austen left by fans. Author's photo

More commercialized touristification aimed at the casual rather than scholarly consumer emerged during the mid-1990s. The trigger was the BBC's adaptation of *Pride and Prejudice*. Visitor numbers at Jane Austen's House Museum had stabilized at about 20,000 people per year by the early 1990s, but *Pride and Prejudice* 'changed everything' (Smith, 2016). Around 55,000 people visited in 1995, and the museum was so crowded that time tickets had to be introduced. After the boom, visitor numbers returned to 37,000–40,000 per year, but this was still almost double the pre-drama levels. The various 200th anniversaries between 2009 and 2017 (Austen moving to Chawton 1809, publication of *Pride and Prejudice* 1813, death in 1817) ensured that visitor numbers moved back towards 50,000 a year.

Jane Austen's House Museum was not the only place seeing increases in visitor numbers. Lyme Park, a stately home in Cheshire, had no connection to Austen prior to the BBC's adaptation, but its visitor numbers soared after the series. The external Pemberley scenes were shot there and Lyme Park became an important destination for Austen fans. Visitor numbers tripled in the year after the broadcast of the series from 32,852 pre-transmission to 91,437 post-transmission (Parry, 2008: 116). Lyme Park featured as a case study in a report commissioned by VisitEngland, *Quantifying Film and Television Tourism in England*. A visitor survey ($n = 193$) on 29 August 2014 determined that 41.3% of visitors cited *screen* productions as a reason for their visit, of whom 90.5% named *Pride and Prejudice* as the work (Olsberg, 2015: 70–72). Visitor numbers in 2015 were 146,675 (VisitEngland, n.d.), which suggests a substantial long-term boost to visitation caused by the drama. This is clearly Austen-related tourism, but cannot be called literary tourism because the iconic 'wet-shirted Darcy' scene, which fixed Lyme Park as Pemberley in the public imagination, was an invention for the drama and is not in the novel at all (Figure 1.2).

Following the explosion of popular interest in Austen triggered by the 1995 screen adaptations, touristification advanced to fulfill the increasing demand from Austen's new fans. The Jane Austen Society of North America (JASNA), set up in 1979, now has over 5000 members and its website lists the tours to England it has offered its members since 1997 (JASNA, n.d.). Meanwhile, Bath also developed its Austen-related tourism. Austen lived in Bath from 1801 to 1806 and it was the setting for both *Northanger Abbey* and *Persuasion*. The city's Georgian architecture makes the city evocative of Austen's era and not simply a place directly related to her life and the novels. The Jane Austen Centre, which is a privately run attraction about her life rather than a museum preserving artefacts, was opened in 1999. In the early 2000s it attracted around 40,000 visitors a year (Kennedy-McLuckie, 2008: 55), but by 2017 the figure had reached 150,000 (Fahy, 2018). Then, in 2001 the first Jane Austen Festival was held in Bath. During the 10 day Festival, there are many events including lectures, public readings, walking tours, musical and stage

Figure 1.2 The author by the famous lake at Pemberley/Lyme Park

performances, and costumed balls. However, the highlight is the Regency Costumed Promenade, when around 500 people in Regency dress walk through the city streets (Figure 1.3). 'Dressing up' in period costume is effectively the same fan behaviour as cosplay (see Rastati, Chapter 9) and as Takayoshi Yamamura argued in his Introduction, cosplay is an important indication of contentsization because fans are reconstructing the narrative world through on-site performance at a tourist event.

Figure 1.3 The Regency Costumed Promenade at the Jane Austen Festival in Bath, 9 September 2017. Author's photo

Table 1.1 Sites of Austen-related contents tourism

Category	Description	Examples
Sites of Austen's life	Her birthplace	Steventon, Hampshire
	Homes	Jane Austen's House Museum, Chawton The houses where she lived in Bath
	Associated Places	Oxford (where her brothers lived and Austen received schooling)
	Death	Grave in Winchester Cathedral
Sites of Austen's era	Regency era architecture	Bath, London
	Napoleonic wars sites	Portsmouth (*HMS Victory*)
	Rural England	The Hampshire countryside
	Country houses	National Trust properties
Novel settings	Real: places described in the novels	Bath (*Northanger Abbey*), Lyme Regis (*Persuasion*), Box Hill (*Emma*)
	Imagined: places assumed to have inspired fictional locations in the novels	Chatsworth House and Kedleston Hall (believed to be inspiration for Pemberley)
Screen adaptation locations	Places where screen adaptations were made	Lacock and Lyme Park (BBC series *Pride and Prejudice*)
Spin-off locations	Places associated with spin-off novels and films	West Wycombe Park (*Austenland*), Basildon Park (*Pride and Prejudice and Zombies*)
Monuments to Austen	Statues	Austen Statue in Basingstoke
	Libraries holding memorabilia	British Library, Bodleian Library, Morgan Library
	Exhibitions (temporary)	Bank of England (2017)
	Attractions	Jane Austen Centre, Bath
Events	Festivals	Jane Austen Festival, Bath
	Performances	Jane Austen the Musical
	Conventions	Meetings of the Jane Austen Society of North America (JASNA)

Other tourist destinations have found it beneficial to associate themselves with Austen. Lyme Regis in Dorset (site of the famous scene in *Persuasion*) and Portsmouth (where *HMS Victory* is moored – Austen's brothers in the navy were involved in the Napoleonic Wars) are the two main examples. There are various tour guidebooks available, including a guide to the filming locations of the various Austen adaptations (Kennedy-McLuckie, 2008) and a walking guide of Austen-related sites in London (Allen, 2013). In the bicentenary of her death in 2017, the Bank of England held a special exhibition to commemorate its new 10 pound note featuring the image of Jane Austen. And in Basingstoke, the nearest town to Austen's birthplace Steventon, which has always taken a lower-key approach to Austen tourism than Bath, there were multiple events including the Sitting

With Jane trail of book benches throughout the town (seats in the shape of an open book decorated with Austen-related images), an exhibition at the Willis Museum & Sainsbury Gallery, and the unveiling of a statue in the town centre. On 17 August 2017, the headline on the front page of the *Basingstoke Gazette* trumpeted, 'The Jane Effect: Basingstoke economy gets a massive boost as Jane Austen fans flock to town in bicentenary year', and a subsequent report by Tourism South East estimated that the Jane Austen 200 commemorations had contributed £21 million to the Hampshire economy (Jane Austen 200, 2018).

All of these incidences of Austen-related touristification reveal that the modern Austen phenomenon has taken tourism well beyond simply the literary and into the realm of mixed media contents tourism. The wide range of sites that now come within the realm of contemporary Austen-related contents tourism are indicated in Table 1.1.

Austen's American Fans in England

Austen's fandom these days is truly global. In 2016, the Marketing Director of Jane Austen's House Museum, Madelaine Smith, told me that, judging by the comments left in visitors' books, there were only three countries worldwide from which the museum had not received a visitor (Smith, 2016). However, within the global fandom, Americans have a particularly important position. According to Herbert's research in 1993–1994, Americans were the second largest national group visiting the Jane Austen House Museum after UK residents (Herbert, 2001: 321). JASNA is the largest Jane Austen fan organization worldwide, and many of the scholarly works, derivative works and travelogues relating to Austen emanate from the USA.

In this final section I focus on contents tourism by Austen's American fans. However, first, what is the broader appeal of Austen in America? The first Austen novel printed in America was an unauthorized reprint of *Emma* released in 1816. Later, the other five novels were published in 1832–1833 as part of a complete edition of her works (Wells, 2017: 25–27). Thereafter, Austen's popularity grew slowly throughout the 19th century, although she had many detractors (such as Mark Twain) as well as admirers.

As Mary Favret has noted, Austen's subsequent popularity in America is somewhat ironic.

> From 1775 to 1817, the span of her life coincides almost exactly with the United States' divorce from England and its efforts to forge an independent national identity [...]. During Austen's lifetime, two wars were fought between the newly hostile nations, the latter of which could have involved one of her naval officer brothers. We might be tempted to say that her novels represent the values of a political system Americans had violently rejected. (Favret, 2000: 166–167)

However, Austen's literary supporters in the USA appreciated the modernity of her novels, a sense that she was 'ahead of her times', and even saw in her writings 'a revolt against the arrogance of rank'.[1] Ultimately, Favret argues the origin of Austen's popularity lies in what may be termed the 'freedom and the pursuit of happiness' (Favret, 2000: 168–178). Austen's following, therefore, was rooted in literary appreciation and self-identification as American before the current trend of Anglophilia and Austen pilgrimages emerged. In its modern guise, Austen's popularity in the USA rests on a wide-ranging set of factors: a longing for an idealized foreign-yet-familiar English rural idyll; a nostalgic view of the English past (often infused with the American penchant for tracing family roots back to the 'Old World'); an image of Austen as a cool feminist ahead of her times whose witty quotes adorn tote bags and provide wise counsel in times of uncertainty; appreciation of Austen in simple literary terms as an author of engaging stories who had mastered the techniques of character development via realistic dialogue; and the author who gave her readers a fantasy of meeting the ultimate English dreamboat, Mr Darcy (especially when imagined as actor Colin Firth). Claudia Johnson (2011: 245) concludes that '"Jane" is "theirs", "yours" and "ours"', in other words, Austen's world provides something for everyone from the serious literary critic to the fan of a spin-off novel/film like *Bridget Jones's Diary*.

Ultimately, all American travellers visiting Austen-related sites in England have their own complex, personal reasons for wanting to deepen their connection with Austen via travel. The relative prominence of the Austen factor in deciding to visit England also varies significantly. The strongest motivations are witnessed in Austen pilgrimages, on which visiting Austen-related sites is the primary motivation or even sole reason for visiting England. The archetypal pilgrimage is the tour organized by the JASNA. The tours take place in July and usually include a ceremony at her grave in Winchester Cathedral on the anniversary of her death (18 July). Deborah Yaffe describes her participation in the 2011 JASNA tour in *Among the Janeites*. Through in-depth portraits of the participants, Yaffe introduces their connections to Austen: the avid novel-reader since childhood, the compiler of a Regency-era website, the 50-year-old who one day discovered *Pride and Prejudice* and Colin Firth as Darcy (Yaffe, 2013: 20–23). She also depicts the demographic profile of the Janeite: 'The thirty-two of us are a pretty homogeneous lot: all white, all American, most female, most middle-aged. Many of us are professionals, active or retired [...]. Four men are loyally escorting their Janeite partners' (Yaffe, 2013: 19–20). This was broadly the demographic makeup of the 52 participants on the 2017 tours, too, judging by the group photo in the tour report in the magazine *Jane Austen's Regency World* (Cooper *et al.*, 2017: 55). These accounts of the JASNA tour confirm that filming locations feature just as prominently in the itineraries as literary locations, although sites directly related to the author such as Chawton and Winchester

Cathedral hold special significance. Furthermore, the camaraderie among like-minded participants is a key reason why they enjoy the trips.

In addition to group trips, there are solo pilgrimages. These tend to be journeys of self-discovery and to develop personal relationships, either an imagined connection with Austen herself or a real connection with a potential Mr Darcy. Accounts of such trips appear on blogs or private social media feeds, but are also published. *A Walk with Jane Austen* by Lori Smith (2007) describes one such pilgrimage. She combines descriptions of a spiritual journey (both religious and overcoming debilitating illness), a search for her Mr Darcy and a travelogue of her month-long trip around Austen-related sites. Like the JASNA trips, the itinerary clearly indicates contents tourism to a range of sites in Austen's world (Table 1.1) rather than something that may be categorized solely as literary tourism. Smith's account is also a notable example of Austen's position as wise counsellor for some fans: throughout this book Austen's works vie with the Bible for the position of Smith's 'life Bible'.

The personal Austen pilgrimage has even featured as the subject of the novel *Austenland* by Shannon Hale (2007), which was made into a film in 2013. In this story, Jane Hayes, an American woman in her thirties obsessed with Colin Firth's Darcy, visits an Austen-themed resort in England in search of romance. The film is both a story about contents tourism because the heroine visits England as a contents tourist, and a story generating contents tourism at its shooting location, West Wycombe Park (for another example of contents tourism in a film generating contents tourism, see the example of *Hello Stranger* in Chapter 10). West Wycombe Park was also a shooting location for the BBC's adaptation of *Sense and Sensibility* (2008) and the mash-up *Pride and Prejudice and Zombies* (2013). This all constitutes 'contentsization' that expands the Austen narrative world with tourism, rather than simply further adaptations, playing a significant role in expanding the contents. Furthermore, unravelling this interconnectedness between scenes and sites within Austen's world may even enhance fans' broader enjoyment of the travel, and is a key indication that the fandom has moved beyond appreciating just the novels and appreciates more broadly Austen-related contents.

All of the pilgrimages described thus far have been published accounts of Austen-related tourism when it was the primary motivation for the international travel experience. However, many people visit Austen-related sites as one part of a general visit to the UK. These travellers spend one or two days on Austen-related attractions and the rest of the time doing other things. To uncover the experiences of this category of traveller, I asked Phil Howe of Hidden Britain Tours for his help in distributing a questionnaire to his clients on his Jane Austen Tour during 2017.[2] This personalized tour goes for a full or half day from Basingstoke through the Hampshire countryside to her birthplace (Figure 1.4), the church in Steventon where her father was rector, Jane Austen's House Museum,

Figure 1.4 The house where Jane Austen was born stood in this field to the right-hand side of the picture. One of the most significant birthplaces in English literature is unmarked and protected from touristification by the local community. Author's photo

the Chawton House Library and the grave of Jane's sister Cassandra. Responses were received by email from 13 tour participants. All were female, international visitors. Eleven were American and the other two were from Australia and Turkey. Their ages ranged from the twenties to seventies, with respondents in their sixties the largest category (five respondents). While not a large sample, each respondent wrote detailed answers running to a page or more of typed A4 that provided insights into these fans and their travel.

Nine respondents mentioned first encountering Austen's novels in their teenage years. Some became immediate, lifelong fans – 'I have been a fan of Jane Austen since I read her books in high school' (Respondent L, seventies) – although some only deepened their interest much later in life. Respondent C (in her fifties) had only first read Austen using a free version of *Emma* on Kindle the previous year. However, screen adaptations and spin-offs also led some respondents to Austen. 'I began reading her books in the 1990s with the release of the *Sense and Sensibility* film with Emma Thompson', said Respondent F (in her sixties). And Respondent J, in her fifties, explained:

> I had minimal exposure to Jane Austen in high school. I rediscovered Jane Austen as an adult with the popularity of the book/movie *Bridget Jones's Diary*. Not only was that story a modern adaptation of *Pride & Prejudice*, but in the book the author specifically mentions the Colin Firth version

of the movie *Pride & Prejudice*, which then introduced me to that movie and reintroduced me to the original *P&P* book (as the lengthy movie is a fairly faithful adaptation of the original book). In turn, that lead me to read other Jane Austen books and seek out other Jane Austen movies, or modern adaptations of her stories. I now have a collection of Jane Austen movies, from Bollywood to British, original to modern versions.

Respondents gave diverse reasons for being Austen fans. The five main reasons were: (a) the depth of characters/relationships and believable scenarios; (b) her woman's perspective ('Jane Austen writes from a woman's heart even across the centuries' – Respondent F, fifties); (c) love stories with happy endings; (d) wit, either Austen's own or that of her characters ('I think mostly I love Elizabeth Bennet – smart, witty, confident – it's wonderful to see a heroine like that in a novel' – Respondent I, 40s); and (e) the lack of profanity, sex and violence ('they are clean and appropriate love stories which are hard to find in our culture today' – Respondent M, thirties).

All of the respondents, therefore, were united in their passion for Austen and their willingness to pay for a personalized tour of related sites. However, Austen-related sites were not the only destinations during their stay in England. Five stated that it was their top priority for their trip to England. For example, Respondent M said:

> We took this tour because our England trip (from the US) was to celebrate my mom's 60th birthday and my mother is the biggest JA fan. This was one of the top things on her list to do while in England.

Those for whom the trip was either a high priority or medium priority (none said it was a low priority) cited either family reasons or other sites that were the primary motivations for the international travel. Respondent L said:

> We were meeting our family in Italy and England was a brief stopover to get over jet lag before joining our family. For me it was a high priority and really the reason for stopping in England rather than going directly to Italy.

And Respondent G said:

> Husband had tickets to 3 Eric Clapton concerts in London and I tagged along since it had been several years since I had been in London. The trip itself into the countryside was a medium priority – Jane was an added bonus. It turned out to be one of the high points of my trip.

The regular mentions of travel partners indicated that within groups or couples there was typically at least one person with high motivation to visit Austen-related sites, while other group members were often just 'tagging along'. Respondent F said, 'It was a medium priority because my travelling partner, my sister, never read any of Jane Austen's books and maybe saw one of the movies. I'm the one with the interest and she was kind enough to indulge me.'

The responses regarding past and future travel also indicated the power of a set of contents, such as Jane Austen, to generate repeat visitation. Of the 13 respondents, five had visited Austen sites in England before. All of these five had visited Bath, one had been to Chawton/Winchester (so the Jane Austen Tour was her second time in Chawton itself) and one had visited the filming locations at Lyme Park and Chatsworth House. Two respondents who had not visited sites in England described attendance at Austen events in America. Many expressed a desire to return again, although plans mostly remained aspirational rather than concrete given the expense involved in getting to England.

Finally, the responses revealed the broader appeal of the UK as a rich destination for contents tourists. In this sense, the whole of the UK may be considered a 'contents brandscape', using the concept proposed by Maree Thyne and Gretchen Larsen in Chapter 2 (see especially Figure 2.1), where an inherently attractive destination is given additional narrative quality by multiple sets of appealing contents co-existing within the same space. For example, Respondent E, in her twenties, gave a detailed account of her England holiday focused on three sets of contents:

> While in England we stayed in London for eight days and Bath for three days. In terms of Jane Austen, we went to Winchester to see Winchester Cathedral, where she is buried, and visited 8 College Street where she died. I went to the exhibit that was held in the Winchester Discovery Centre where they had her Pelisse, portraits, and first edition novels. We also went to Bath for Jane. We visited the Jane Austen Centre, took a Jane Austen walking tour around the city, visited the Assembly Rooms where she went to public balls, and saw where she lived at 4 Sydney Place. We also walked around Sydney Gardens where she strolled through in the early 1800s. Visiting Jane Austen sites while in England was a high priority. The reason it was not top priority was because I also traveled to London to see Harry Potter sites (*Harry Potter and the Cursed Child* play, King's Cross Station and Platform 9¾, Warner Bros Studio Tour London, and Harry Potter filming locations). Both Jane Austen and Harry Potter were high priorities while visiting England as well as seeing Highclere Castle (*Downton Abbey*) and the London attractions.

According to Jane Austen Tour guide Phil Howe (2018), he has many clients who mention an interest in Charles Dickens, the Brontë sisters, Beatrix Potter and (particularly those with younger families) Harry Potter as well as Jane Austen. Many of the points that make Austen tourism attractive – popular novels, film adaptations, an appealing narrative world and a wide range of related (and appealing in their own right) heritage sites to visit – also make Harry Potter tourism a significant form of inbound contents tourism. I will return to this point in the Conclusions chapter of this book.

Conclusions

In this chapter I have identified 1995 as a watershed in Austen-related tourism. The popularity of the BBC *Pride and Prejudice* adaptation, along with two other major screen adaptations, ignited a broader interest in Jane Austen that was clearly reflected in increased tourism numbers to related sites (Jane Austen's House Museum and Lyme Park), and the subsequent development of new tourism sites (Jane Austen Centre), events (Bath Jane Austen Festival) and tourism products (JASNA tours and the Jane Austen Tour[3]). As the influence of screen adaptations on Austen's world has increased, it has become less appropriate to talk about Austen-related tourism as literary tourism. Spin-off novels promote further contentsization, and when filmed (such as *Austenland*, *Pride and Prejudice and Zombies* and *Death Comes to Pemberley*) can even generate tourism to sites that had already been connected to Austen's world by more faithful screen adaptations of Austen novels. Driving this contentsization are not the literary aficionados, but rather the 'Austen omnivores' (Wells, 2011: Chapter 2), who consume anything Jane-related, and 'amateur readers' (Wells, 2011: Chapter 3), whose primary purpose is enjoyment not erudition. These are the fans who make the contemporary Austen industry commercially viable and tourism is central to both their fan activities and the ongoing development of 'Austen's world'.

America's role in the current Austen boom is significant. Many of the screen adaptations are produced in America or with American financial backing, and many of the international tourists visiting Austen-related sites are American. At the same time, surveys of those American visitors or accounts of travel experiences written by them reveal diverse interests while travelling in England, and even those on pilgrimages can have other motivations too. On JASNA group tours, companionship with like-minded travel partners is clearly significant. However, there are also many people 'tagging along' at Austen-related sites. These people, who might appear at first to be contents tourists, are actually motivated by a desire to maintain a relationship with the travel partner, and not necessarily to deepen a relationship with the contents.

The survey of Hidden Britain Tours participants also reveals that many people are interested in contents tourism as a general *type* of tourist experience. In other words, they enjoy visiting places given extra narrative quality or significance by association with works of popular culture. These travellers enjoy multiple sets of contents, which is why the Jane Austen fans in the survey said they also wanted to visit sites related to Thomas Hardy or Harry Potter on future trips, rather than be limited to tourist experiences within one narrative world. Contents tourists in this view are not so much fans of a specific set of contents (author, character or franchise), but fans of a particular style of tourism experience which connects their present travel experience to their enjoyment of past and future consumption of mediatized culture.

As the narrative world based on Jane Austen is developed even further by more and more derivative works, screen adaptations and reinterpretations of her work – whether faithful to the original or mash-up, scholarly or purely for fun – we can expect the meanings and forms of Austen-related travel to diversify ever further. The shift in Austen-related tourism, therefore, can be expected to continue on its current trajectory away from a more traditional and narrowly defined form of literary tourism towards the more complex and multifaceted form of contents tourism.

Notes

(1) Here Favret (2000: 184, n. 14) is citing comments by William Dean Howells published in *Harper's Bazaar* in 1900.
(2) I took the Hidden Britain Tour on 6 September 2016. I am very grateful to Phil Howe for his support of the research project, for distributing the questionnaire to clients, and for his helpful comments on the first draft of this paper. My gratitude, also, to the respondents who provided such detailed information in their responses.
(3) In *A Walk with Jane Austen*, Lori Smith describes an encounter with Phil Howe in 2006, who was developing his Jane Austen Tour at the time (Smith, 2007: 102).

References

Allen, L. (2013) *Walking Jane Austen's London: A Tour Guide for the Modern Traveler*. Oxford: Shire.
Beeton, S., Yamamura, T. and Seaton, P. (2013) The mediatisation of culture: Japanese contents tourism and popular culture. In J.-A. Lester and C. Scarles (eds) *Mediating the Tourist Experience: From Brochures to Virtual Encounters* (pp. 139–154). Farnham: Ashgate.
Chronis, A. (2012) Between place and story: Gettysburg as tourism imaginary. *Annals of Tourism Research* 39 (4), 1797–1816.
Cooper, L.P., Rosowicz, J. and Moss, C.M. (2017) Touring with Jane. *Jane Austen's Regency World* November/December, 54–55.
Crang, M. (2003) Placing Jane Austen, displacing England: Touring between book, history and nation. In S.R. Pucci and J. Thompson (eds) *Jane Austen and Co.: Remaking the Past in Contemporary Culture* (pp. 111–130). New York: State University of New York Press.
Fahy, V. (2018) Personal correspondence with Victoria Fahy (Visitor Experience Manager, Jane Austen Centre, Bath), 6 June.
Favret, M.A. (2000) Free and happy: Jane Austen in America. In D. Lynch (ed.) *Janeites: Austen's Disciples and Devotees* (pp. 166–187). Princeton, NJ: Princeton University Press.
Hale, S. (2007) *Austenland*. New York: Bloomsbury.
Harman, C. (2009) *Jane's Fame: How Jane Austen Conquered the World*. New York: Picador.
Herbert, D. (2001) Literary places, tourism and the heritage experience. *Annals of Tourism Research* 28 (2), 312–333.
Higson, A. (2011) *Film England: Culturally English Filmmaking since the 1990s*. London: I.B. Tauris.
Hills, M. (2015) From 'Multiverse' to 'Abramsverse': *Blade Runner, Star Trek*, multiplicity, and the authorizing of Cult/SF worlds. In J.P. Telotte and G. Duchovnay (eds) *Science Fiction Double Feature: The Science Fiction Film as Cult Text* (pp. 21–37). Liverpool: Liverpool University Press.

Howe, P. (2018) Personal correspondence with Phil Howe (Hidden Britain Tours), 29 June.
Jane Austen 200 (2018) The Jane effect: Jane Austen 200 gives Hampshire a £21 million boost. See http://janeausten200.co.uk/jane-effect-jane-austen-200-gives-hampshire-£21-million-boost (accessed June 2018).
JASNA (n.d.) Tours of England. See http://www.jasna.org/conferences-events/tour/ (accessed June 2018).
Jenkins, H. (2006) *Convergence Culture: Where Old and New Media Collide*. New York: New York University Press.
Johnson, C.L. (2011) Austen cults and cultures. In E. Copeland and J. McMaster (eds) *The Cambridge Companion to Jane Austen* (2nd edn) (pp. 232–247). Cambridge: Cambridge University Press.
Jones, H. (2014) *Jane Austen's Journeys*. London: Robert Hale.
Kennedy-McLuckie, M. (2008) *Jane Austen: TV & Film Locations Guide*. Durham, UK: Trail.
Olsberg SPI (2015) *Quantifying Film and Television Tourism in England*. See http://applications.creativeengland.co.uk/assets/public/resource/140.pdf (accessed June 2018).
Parry, S. (2008) The Pemberley effect: Austen's legacy to the historic house industry. *Persuasions* 30, 113–122.
Reijnders, S. (2011) *Places of the Imagination: Media, Tourism, Culture*. Farnham: Ashgate.
Seaton, P., Yamamura, T., Sugawa-Shimada, A. and Jang, K. (2017) *Contents Tourism in Japan: Pilgrimages to 'Sacred Sites' of Popular Culture*. New York: Cambria Press.
Smith, L. (2007) *A Walk with Jane Austen*. Colorado Springs, CO: Lion.
Smith, M. (2016) Interview conducted by the author at Jane Austen House Museum, 6 September.
Sutherland, K. (2011) Jane Austen on screen. In E. Copeland and J. McMaster (eds) *The Cambridge Companion to Jane Austen* (2nd edn) (pp. 215–231). Cambridge: Cambridge University Press.
Todd, J. (2015) *The Cambridge Introduction to Jane Austen*. Cambridge: Cambridge University Press.
VisitEngland (n.d.) Most visited paid attractions – North West 2015. See https://www.visitbritain.org/sites/default/files/vb-corporate/Documents-Library/documents/England-documents/most_visited_paid_north_west_2015b.pdf (accessed May 2018).
Wells, J. (2011) *Everybody's Jane: Austen in the Popular Imagination*. London: Bloomsbury.
Wells, J. (2017). Austen in America. *Jane Austen's Regency World*. November/December, 24–29.
Yaffe, D. (2013). *Among the Janeites: A Journey Through the World of Jane Austen Fandom*. Boston, MA: Mariner Books.

2 Conceptualizing Contents Brandscapes: The Brontë Brand

Maree Thyne and Gretchen Larsen

Introduction

The Brontës are revered throughout the world for their literary prowess, and undoubtedly the Brontë brand has a very strong international following which attracts many fans to seek out and visit the home of these authors. The small village of Haworth, and its surroundings of West Yorkshire, England, are inextricably linked with the Brontës. Not only did they live, work and write there, but the place frequently intersects their writing as both a setting and an influence on the characters within their novels. Tourists now flock to experience these different renderings of Haworth. While in the 1850s there was only a 'trickle of visitors to Haworth' (Barker, 1995: 830), tourism is now a mainstay industry for the village and its surrounding community, and every year thousands of visitors come to the area which 'housed such febrile talent' (Barker, 1995: 830).

Related studies into the Brontës have previously focused on the link between the components of the Brontë brand and literary tourism (Barnard, 2002; Hoppen *et al*., 2014; Tetley & Bramwell, 2002), emphasizing the connections which visitors to Haworth have to the authors or the texts. However, in a similar vein to Seaton's discussion of Austen-related tourism in Chapter 1 of this book, we investigate the multiple narrative worlds of the Brontë brand – made up of (multiple) authors, their characters, novel settings, film locations and songs. Taken individually, these elements can induce and attract specific tourism industries, such as literary tourism and film tourism, but together as a narrative world, they inspire contents tourism: 'travel behaviour motivated fully or partially by narratives, characters, locations and other creative elements of popular culture forms, including film, television dramas, manga, anime, novels and computer games' (Seaton *et al*., 2017: 263). In comparison with Austen-related tourism, the multiple narrative worlds of the Brontës are collectively set in one small area of Yorkshire in northern England and are combined and used as marketing and branding tools in promoting Haworth and the Brontë Country. The *physical and natural world* of Haworth and the Brontë Country and the *narrative worlds* provided by the Brontës are inextricably linked in a complex landscape of interconnected and interrelated brands – arguably creating two related brandscapes: a contents brandscape and a place brandscape.

This chapter argues the necessity to understand visitor experiences, expectations and, in particular, how visitors manage their experience when a contents tourism destination comprises multiple elements and brands, ranging from real to fictitious, and varying from birthplaces to places of inspiration, story settings, homes, schools and graveyards. However, in order to garner meaningful insight into the contents tourist, we first need to understand the intersections, interrelationships, tensions and alignments that exist between the art form and the place within a contents brand.

There are many components which make up a contents brand, of which the place itself also needs to be considered. This is certainly the case for the Brontë brand, which comprises a number of different (although connected) brands, and place is a vital thread throughout. We adopt and adapt the notion of a 'brandscape' (Sherry, 1998; Thompson & Arsel, 2004) as a 'loosely bounded site within which meaning is derived from making sense of the various, interrelated brands' (O'Reilly & Kerrigan, 2013: 769) to the context of contents brands, as a conceptual framework to understand how various brands work together to form the contents brandscape and how it connects to the place brandscape within the Brontë country. After locating the notion of a contents brandscape in relation to the existing literature on contents tourism and brandscapes, we present a model of a contents brandscape, applying it to the case of the Brontës and analysing the connections with the place brandscape.

Contents (from the Japanese term *kontentsu*) is a 'combination of creative elements' (Seaton *et al.*, 2017: 2); contents brands are made up of multiple components, of which art and place are just a few. Contents include everything from narratives, characters, locations and producers, through to other creative elements (such as music) within works of mediatized popular culture such as novels, film and television (Seaton *et al.*, 2017). Therefore, contents tourism is complex, multi-layered and interconnected. To better understand how various brands relate to one another in the context of contents brands, without prioritizing one particular type of brand (e.g. place or art), we adopt and adapt the notion of a 'brandscape'. Approaching contents brands through a brandscape lens enables us to unmask their complexity and therefore to understand how consumers and other stakeholders experience, and make sense of, brands that exist in relation to one another.

Brandscapes

Introduced to marketing and consumer research by Sherry (1998), a brandscape generally refers to a 'loosely bounded site within which meaning is derived from making sense of the various, interrelated brands' (O'Reilly & Kerrigan, 2013: 769). Previous studies have explored brandscape in relation to Starbucks as a glocalized and hegemonic brandscape (Thompson & Arsel, 2004) and in contexts where multiple brands and stakeholders coalesce around each other, such as in film (O'Reilly & Kerrigan, 2013). This chapter argues for the extension and adaptation of O'Reilly and Kerrigan's (2013) notion of the film brandscape into the field of contents brands.

Considering contents brands and tourism through the lens of the brandscape highlights three important issues. Firstly, as a contraction of 'branded landscapes', the notion of a brandscape calls attention to the changed nature of culture and place from those characterized by village life and nature, to that which is dominated by commerce and brands. Salzer-Mörling and Strannegård (2010: 412) define a brandscape as 'a culture or a market where brands and brand-related items such as signs and logos increasingly dominate everyday life'. A consequence of the experience economy's focus on the expressive, sign values of goods, the marketplace has thus become 'a battlefield of brand names, images and logos striving to be heard' (Christensen & Cheney, 2000: 247). This is a particularly useful understanding in the case of destinations, which can be conceived of as places transformed through marketing and commercial practice into a 'landscape of impression and immersion' (Salzer-Mörling & Strannegård, 2010: 231) that is designed to facilitate visitor experiences exclusively associated with that place brand (Ponsonby-McCabe & Boyle, 2006).

Secondly, the notion of brandscapes underscores the relational and temporal nature of brand-related meaning and the agency of the consumer in constructing 'personal meanings and lifestyle orientations from the symbolic resources provided by an array of brands' (Sherry, 1998: 112). The notion that brands are consumed in relation to other brands is widely acknowledged both in the arts (e.g. Preece & Kerrigan, 2015) and in place branding (e.g. Kavaratzis & Kalandides, 2015) as the value associated with these kinds of products is experiential and symbolic. Meaning comes not only from marketing communications and branding strategies, but from previous experience with the brand or related brands, and a multitude of other cultural cues. Thus, meanings are developed and circulated in a continuous reciprocal interplay within the brandscape (O'Reilly & Kerrigan, 2013; Närvänen & Goulding, 2016).

Following on from this, brandscapes highlight the interconnectedness of the brands. This can be visualized as a network which then can be mapped out, as done by O'Reilly and Kerrigan (2013) in the context of film: 'where, because of the multiple stakeholder involvement [...], a number of brands co-exist in the same project, for example, people brands (director, actor, screenwriter) and commercial brands (product placements)' (O'Reilly & Kerrigan, 2013: 773). Approaching contents brands through brandscape introduces the idea of granularity and also nuance in teasing out the inter-relationships, alignments, conflicts and tensions between and within brands that almost certainly exist in such a setting (Brown, 2011), which various stakeholders need to negotiate.

The Relationship between Contents Brands and Place Brands

All kinds of places, from countries and states, to cities and towns, brand themselves in order to draw visitors and/or encourage people to relocate there. Virtually every physical location, area or region considers place branding. Place branding encompasses the broad set of 'efforts by country, regional and city governments, and by industry groups, aimed at marketing the places and sectors they represent' (Papadopoulos, 2004: 36). When applied to tourism, the goal is to communicate a destination's special identity and uniqueness by differentiating its character and offering from that of competing destinations (Qu *et al.*, 2011). Lee *et al.* (2015) argue that this should involve the identification of the appealing characteristics of place and the experiential tourism activities on offer. One way this might be achieved is by associating a place with a recognized artist, in the hope that 'the necessary unique qualities of the individual are transferred by association to the place' (Ashworth, 2009: 11). While Ashworth (2009) offers examples of figures who can be used in this way – such as artists (Gaudí in Barcelona), musicians (Mozart in Salzburg) and writers (the Brontës in Yorkshire) – the growing area of contents tourism is:

> better suited to explaining the travel behaviors of fans who have a strong interest in a particular set of contents or 'world' – comprising narratives, characters, locations and other creative elements – that is created by franchises, artists, genres, and contents businesses and is disseminated across various media formats. (Seaton *et al.*, 2017: 265)

Although research into contents and associated travel and travel experiences has been relatively sparse, studies are increasing. However most research has analysed individualized associates of contents tourism such as film-induced tourism (Beeton, 2016; Macionis & Sparks, 2009), literary tourism (Hede & Thyne, 2010; Herbert, 2001; Seaton, 1999; Squire, 1994; Young, 2015) and music-induced tourism (Gibson & Connell, 2007; Kay, 2006) – rather than exploring the 'accelerated multiuse of storylines, characters, locations and other creative elements across various media formats' (Seaton *et al.*, 2017: 2), which a contents tourism lens allows us to do. Not only that, but by using the Brontë Country as an example, we have the pull of not just one but three or four associated artists.

Currently place branding strategies have focused on close associates of contents tourism, for example there are many examples where the relationship between literature and place has been used as a key ingredient of a place brand, whether this be the setting within a novel (The Rosslyn Chapel in *The Da Vinci Code* by Dan Brown; Monterey, California in *Cannery Row* by John Steinbeck), the birthplace of an author (Shakespeare's Stratford-upon-Avon), the award of UNESCO's city of literature (Dublin, 2010; Edinburgh, 2004), designation as a national book town (Wigtown, Scotland) or hosting a literary festival (Hay-on-Wye). While there remain many aspects of literary tourism that are still worthy and in need of examination, the focus here is on contents tourism and understanding the relationship between contents and place. Adopting the lens of 'place brands' as a tool to begin to make sense of the relationship between art and place in contents tourism is useful insofar as it highlights the embeddedness and importance of art in place and vice versa (Butler, 2000). However, the limitation is that it focuses the analytical lens on the place element, and thus masks the scope of, and interconnections between, the different kinds of brands involved. Second, we are also interested in the nature of contents tourism in contexts where more than one artist brand is present (Charlotte, Emily, Anne and arguably Branwell).

A contents tourism industry does not happen on its own, particularly in a time of increased and easier access to a number of related attractions. Even such an established literary destination as Stratford-upon-Avon is driven to organize festivals and special events, and to coordinate rather than compete with a range of other brands within the destination (Barnard, 2002) in order to encourage visitors to Shakespeare country. Shifting our focus from literary tourism to contents tourism provides the platform for developing a more inclusive and integrated understanding of interrelationships amongst these different types of contents that might be associated with a literary place.

The Brontë Brandscape

In adapting O'Reilly and Kerrigan's (2013) film brandscape to the contents context, we extend the parameters beyond a singular type of brand (i.e. the film brandscape) to one that comprises at least two different types of interrelated brandscapes – the place brandscape and the contents brandscape (see Figure 2.1). Within each of these brandscapes, multiple brands can be identified. What is also highlighted in Figure 2.1 is the interrelationship between the individual brands within each brandscape, and between the place and contents brandscapes. Note that the contents brandscape also includes other contents-based narrative worlds, which are set in or near Haworth. While the 'Brontës World' is the main set of contents upon which Haworth is marketed, subsidiary sets of contents

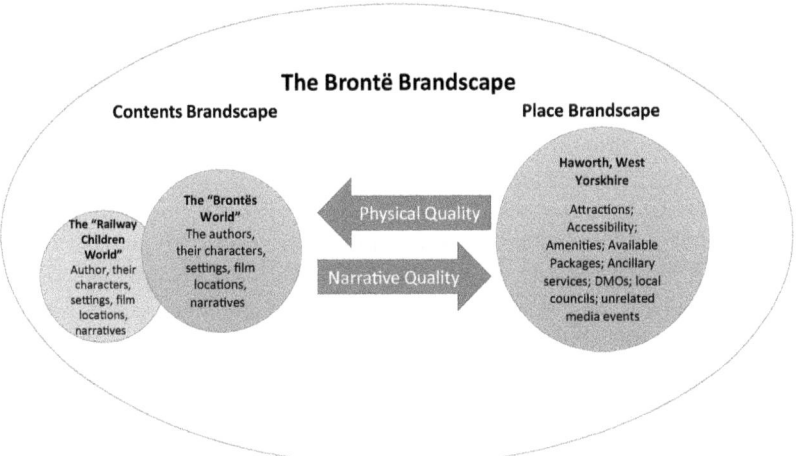

Figure 2.1 The Brontë contents brandscape

(for example, the 'Railway Children World'[1]) add depth, breadth and a more diverse appeal to the collective Brontë brandscape.

The inextricable connection between place and contents is exemplified by the Brontës brandscape. This connection has drawn visitors to Haworth for over 160 years, with some of the first descriptions of Haworth as a destination being provided by Elizabeth Gaskell in her biography of her friend Charlotte Brontë, published in 1857 (only two years after Charlotte's death). Gaskell, herself a frequent traveller, describes the trip between Keighley and Haworth – expecting 'brilliancy and vividness' but instead struck with a 'feeling of disappointment at the grey natural tint of every object, near or far off' (Gaskell, 1857: 3). She goes on to describe the vegetation: 'it does not flourish, it merely exists; and instead of trees there are only bushes and shrubs about the dwellings ... and what crops there are, on the patches of arable land, consist of pale, hungry-looking, grey-green oats'. The now infamous and frequently visited Yorkshire Moors also do not escape the Gaskell wrath:

> sinuous wave-like hills [...] crowned with wild bleak moors – grand from the ideas of solitude and loneliness which they suggest, or oppressive from the feeling which they give of being pent up by some monotonous and illimitable barrier, according to the mood of mind in which the spectator may be. (Gaskell, 1857: 4)

Charlotte Brontë herself expressed surprise in the fact that 'various folks are beginning to come boring to Haworth on the wise errand of seeing the scenery described in Jane Eyre and Shirley' (Smith, 2000: 353). Also published by Smith is a letter which Charlotte wrote to one of her publishers, describing Haworth as 'a strange uncivilized little place' (Smith, 2000: 240).

The image of the Yorkshire Moors that many hold does indeed still reflect a 'wiley windy' place, as evoked by Kate Bush in her 1978 critically acclaimed and enormously successful debut single 'Wuthering Heights', which was inspired by the 1967 BBC adaptation of the novel of the same name. Neither this image, nor the words of Gaskell or even Charlotte Brontë herself, have extinguished the strong desire which thousands of visitors clearly have had for over 160 years to visit the place most closely aligned and associated with the Brontë family and their work. Visitor numbers have waxed and waned over the years, with the highest annual visitor numbers recorded in 1974 at 221,000. Recent visitor figures, although less per annum, appear to be growing, with latest figures of 88,000 per year reflecting a growth of admissions to the Brontë Parsonage Museum of over 20% on the previous year, following an increase in 2016 of nearly 10% from 2015 (The Brontë Society, n.d. *a*).

Barnard (2002) associates a few major influences on enabling and indeed influencing the tourism development of Haworth and its surrounding area. The establishment of the Brontë Society in 1893 has been hugely influential in preserving the legacy of the Brontës. Indeed this society may have been the first literary society dedicated to novelists (Barnard, 2002). The popularity of the early film adaptations of *Wuthering Heights* and *Jane Eyre* arguably sparked further interest in the novels in the early 1900s (much as adaptation of texts do today). However, possibly more importantly to the tourism industry of Haworth was the sale of the old vicarage by the Church of England to the Brontë Society in 1927. 'Before, one came to see the Brontës' home. Now one could visit it', cementing Haworth as a 'place of literary pilgrimage' (Barnard, 2002: 148) and a physical anchor point for related contents tourism.

The Brontë Society continues to breathe life into the Brontë brandscape, consistently offering new insights and perspectives into the family, ways for visitors to experience their work and lives, and collaborative support for the development of new contents. Recent examples include the 2017 exhibit at the Parsonage Museum dedicated to Branwell Brontë. *Mansions in the Sky* was curated by the poet Simon Armitage and arguably set up to 'tone down the Branwell bashing, recognizing his flaws but celebrating the merits of the brother' (Butcher, 2017). The second example also occurred in 2017, when visitors to the Parsonage museum were invited to co-create a handwritten copy of *Wuthering Heights*. During the year, over 10,000 visitors each contributed one handwritten line from the novel into a book which was then exhibited in 2018, the bicentenary of Emily Brontë's birth. Each participant was gifted a pencil, commissioned by the artist Clare Twomey, 'as a tool for further writing'. Twomey added that she hoped that the act of 'sitting at a table in the house where Emily wrote her novel, and to hold a pencil and write, will build understanding of Emily and her determination to create the one published work of her lifetime' (Brontë Society, n.d. *b*). Aside from igniting a connection with

Emily, this experience arguably makes literature more accessible as all visitors were invited to be involved – perhaps a good example of what Gentile and Brown (2015: 27) stress as a major advantage of literary tourism: turning 'language into spectacle [...] making literature and art more accessible to wider audiences'.

To make sense of this complex relationship between place and art within the Brontë brandscape, both the place brandscape and the contents brandscape will now be explained further.

Place brandscape

Zenker and Braun (2010: 4) define a place brand as 'a network of associations in the consumers' mind based on the visual, verbal, and behavioural expression of a place, which is embodied through the aims, communication, values, and the general culture of the place's stakeholders and the overall place design'. Various studies have investigated different facets of the place brand, and many are linked to tourism provisions. One of the widely acknowledged key challenges is that image is critical to a place brand, but that a place brand image is created in people's minds as they encounter the place and therefore cannot be controlled via branding strategies:

> the place brand is based on what a place 'has' 'does' and 'shows' but it is constructed by each individual separately as it refers to a mental process that occurs in the mind of individual people. It is individuals that know and use various associations with the destination, the combination of which creates the brand. Naturally, different individuals will construct the brand differently as they will have different levels of knowledge, understanding, interest, affection and so on. (Kavaratzis, 2017: 97)

Implicit in this is that place brands are complex, especially in comparison with other kinds of brands such as goods and services (Kaplan *et al.*, 2010), and that the place brand is likely to incorporate relationships with a range of other kinds of brands that exist in that place.

This conceptualization resonates with O'Reilly and Kerrigan's (2013) notion of brandscape. We argue that the place brandscape incorporates all those brands involved in the production, consumption and marketing of place and destination. These include Buhalis's (2000) six As of tourism destinations: attractions (natural, man-made, artificial, purpose-built, heritage and special events), accessibility (entire transport system comprising routes, terminals and vehicles), amenities (accommodation and catering facilities, retailing, other tourist services), available packages (pre-arranged packages by intermediaries and principals), activities (all activities available at the destination and what consumers will do during their visit) and ancillary services (services used by tourists such as banks, communications, post, newsagents, hospitals, etc.). We would also add to this the brands involved in managing and marketing the destination including Destination Management Organizations, local and national marketing agencies, travel agencies and local and national authorities.

Contents brandscape

The contents brandscape includes those brands which relate to the narrative world, such as the stories, characters and locations, and other works inspired by the original art. In this case the Brontë brandscape comprises three (or four if you include Branwell Brontë) separate artist brands (Charlotte, Emily and Anne), each of whom produced various works of art and around which further contents have been created (e.g. Armitages's *Mansions in the Sky*). This differs significantly from other important literary-based contents brands (such as William Shakespeare, Beatrix Potter and Jane Austen), where there is a single artist brand at the core. In addition to the artist and their art, the contents brandscape also includes the stories and characters of the novels (e.g. Heathcliff), settings (both imagined (Wuthering Heights) and real (the Yorkshire Moors)), the built environment of the lives of the artists, (e.g. Brontë Parsonage Museum, Haworth School, Haworth village, Haworth Cemetery) and related contents that are based on brand extensions and adaptations (e.g. poems, novels, plays, films, children's books, documentaries, academic texts, museum exhibitions). The contents brandscape also includes other, subsidiary contents brandscapes such as the 'Railway Children World', which are also connected to Haworth to varying degrees.

The contents brandscape is complex and interconnected. Not only do a number of brands exist in the brandscape, but so too do a number of stakeholders. Visitors to Haworth may be drawn to the destination for any number of reasons, reasons which may fit into the contents brandscape and/or the place brandscape. The relevance and contribution of this research is based on the acknowledgement that in order for consumers (visitors) to engage with a brand they view as authentic, the brand story needs to make visible the underlying complexities and tensions that exist within a brand (Beverland, 2009) and arguably a brandscape. This chapter produces a conceptual model illustrating the complexity of a contents brandscape. However in order to understand how visitors negotiate and make sense of a complex brandscape (in the hope of forming positive place brand images), future research needs to include the collection of empirical data with visitors and other stakeholders in and around Haworth. This data would reflect the experiences which stakeholders have of the destination, and how they negotiate tensions drawn from the interconnected brandscape that they find themselves in.

Concluding Comments

In conclusion, a brandscape lens provides a useful framework for understanding how consumers (or any stakeholder) engage with and create meaning around brands in relation to one another, and how they might negotiate this complexity in their experience with the contents brandscape. For example, how does a visitor reconcile their love of Charlotte Brontë with their aversion to Bramwell and his vices? Or their pleasure at visiting the 'wiley windy' moors of their Brontë inspired-imagination which lie alongside West Yorkshire's post-industrial decay? Or how do the experiences of the host community align or conflict with those of visitors? As Baumgarth and O'Reilly (2014: 5) argue of arts brands, but which is equally applicable to all contents brands, there 'is a need to conceive somehow of arts brands being in shared ownership, as a kind of artistic commons (morally, if not financially), or at least something in which many stakeholders have different stakes or interests'. Brandscapes enable such a conception, and a more nuanced and critical understanding of the alignments and tensions that exist among brands and actors in the brandscape.

Note

(1) *The Railway Children*: The narrative world of *The Railway Children* has its origin in the children's book by Edith Nesbit, published in 1906 and based on a 1905 serial in *The London Magazine*. The book tells the story of three children who, following the false imprisonment of their father for spying, are forced to move from London with their mother to the 'Three Chimneys', a fictional house near the railway in Yorkshire. Through the 1970 film adaptation of the novel, the location of this house and railway took a more concrete physical and narrative shape as The Keighley & Worth Valley Railway, a heritage line that runs through Haworth Train Station. Both were used extensively during filming, and various tourist offerings now provide the visitor with the opportunity to step into the world of the Railway Children.

References

Ashworth, G. (2009) The instruments of place branding: How is it done? *European Spatial Research and Policy* 16 (1), 9–22.
Barker, J. (1995) *The Brontës*. London: Weidenfeld & Nicolson.
Barnard, R. (2002) Tourism comes to Haworth. In M. Robinson and H.C. Andersen (eds) *Literature and Tourism* (pp. 143–154). London: Thomson Learning.
Baumgarth, C. and O'Reilly, D. (2014) Brands in the arts and culture sector. *Arts Marketing: An International Journal* 4 (1/2), 2–9.
Beeton, S. (2016) *Film-Induced Tourism* (2nd edn). Bristol: Channel View Publications.
Beverland, M. (2009) *Building Brand Authenticity: 7 Habits of Iconic Brands*. Basingstoke: Palgrave Macmillan.
The Brontë Society (n.d. *a*) Museum celebrates increase in visitor numbers. See https://www.bronte.org.uk/whats-on/news/213/museum-celebrates-increase-in-visitor-numbers (accessed January 2019).
The Brontë Society (n.d. *b*) Clare Twomey: Wuthering Heights – a manuscript. See https://www.bronte.org.uk/whats-on/381/clare-twomey-wuthering-heights-–-a-manuscript/386 (accessed January 2019).

Brown, S. (2011) And then we come to the brand: Academic insights from international bestsellers. *Arts Marketing: An International Journal* 1 (1), 70–86.

Buhalis, D. (2000) Marketing the competitive destination of the future. *Tourism Management* 21 (1), 97–116.

Butcher, E. (2017) It's time to bring Branwell, the dark Brontë, into the light. *The Guardian*, 26 June. See https://www.theguardian.com/books/2017/jun/26/its-time-to-bring-branwell-the-dark-bronte-into-the-light (accessed January 2019).

Butler, P. (2000) By popular demand: Marketing the arts. *Journal of Marketing Management* 16 (4), 343–364.

Christensen, L.T. and Cheney, G. (2000) Self-absorption and self-seduction in the corporate identity game. In M. Schultz, M.J. Hatch and M.H. Larsen (eds) *The Expressive Organization – Linking Identity, Reputation, and the Corporate Brand* (pp. 246–270). Oxford: Oxford University Press.

Gaskell, E. (1857) *The Life of Charlotte Brontë*. Oxford: Oxford University Press.

Gentile, R. and Brown, L. (2015) A life as a work of art: Literary tourists' motivations and experiences at il vittoriale Degli italiani. *European Journal of Tourism, Hospitality and Recreation* 6 (2), 25–47.

Gibson, C. and Connell, J. (2007) Music, tourism and the transformation of Memphis. *Tourism Geographies* 9 (2), 160–190.

Hede, A.-M. and Thyne, M. (2010) A journey to the authentic: Museum visitors and their negotiation of the inauthentic. *Journal of Marketing Management* 26 (7–8), 686–705.

Herbert, D. (2001) Literary places, tourism and the heritage experience. *Annals of Tourism Research* 28 (2), 312–333.

Hoppen, A., Brown, L. and Fyall, A. (2014) Literary tourism: Opportunities and challenges for the marketing and branding of destinations? *Journal of Destination Marketing and Management* 3 (1), 37–47.

Kaplan, M.D., Yurt, O., Guneri, B. and Kurtulus, K. (2010) Branding places: Applying brand personality concept to cities, *European Journal of Marketing* 44 (9/10), 1286–1304.

Kavaratzis, M. (2017) The participatory place branding process for tourism: Linking visitors and residents through the city brand. In N. Bellini and C. Pasquinelli (eds) *Tourism in the City – Towards an Integrative Agenda on Urban Tourism* (pp. 93–107). Cham: Springer.

Kavaratzis, M. and Kalandides, A. (2015) Rethinking the place brand: The interactive formation of place brands and the role of participatory place branding. *Environment and Planning A* 47 (6), 1368–1382.

Kay, A. (2006) Promoting tourism through popular music. *Tourism, Culture & Communication* 6 (3), 209–215.

Lee, A.H.J., Wall, G. and Kovacs, J.F. (2015) Creative food clusters and rural development through place branding: Culinary tourism initiatives in Stratford and Muskoka, Ontario, Canada. *Journal of Rural Studies* 39, 133–144.

Macionis, N. and Sparks, B. (2009) Film-induced tourism: An incidental experience. *Tourism Review International* 13 (2), 93–101.

Närvänen, E. and Goulding, C. (2016) Sociocultural brand revitalization: The role of consumer collectives in bringing brands back to life. *European Journal of Marketing* 50 (7/8), 1521–1546.

O'Reilly, D. and Kerrigan, F. (2013) A view to a brand: Introducing the film brandscape. *European Journal of Marketing* 47 (5/6), 769–789.

Papadopoulos, N. (2004) Place branding: Evolution, meaning and implications. *Place Branding* 1 (1), 36–49.

Ponsonby-Mccabe, S. and Boyle, E. (2006) Understanding brands as experiential spaces: Axiological implications for marketing strategists. *Journal of Strategic Marketing* 14 (2), 175–189.

Preece, C. and Kerrigan, F. (2015) Multi-stakeholder brand narratives: An analysis of the construction of artistic brands. *Journal of Marketing Management* 31 (11–12), 1207–1230.

Qu, H., Kim, L.H. and Im, H.H. (2011) A model of destination branding: Integrating the concepts of the branding and destination image. *Tourism Management* 32 (3), 465–476.

Salzer-Mörling, M. and Strannegård, L. (2004) Silence of the brands. *European Journal of Marketing* 38 (1/2), 224–238.

Seaton, A.V. (1999) Book towns as tourism developments in peripheral areas. *International Journal of Tourism Research* 1 (5), 389–399.

Seaton, P.A., Yamamura, T., Sugawa-Shimada, A. and Jang, K. (2017) *Contents Tourism in Japan: Pilgrimages to 'Sacred Sites' of Popular Culture*. Amherst, NY: Cambria Press.

Sherry, J.F. (1998) The soul of the company store: Nike Town Chicago and the emplaced brandscape. In J. Sherry (ed.) *Servicescapes: The Concept of Place in Contemporary Markets* (pp. 109–146). Lincolnwood, IL: NTC Business Books.

Smith, M. (ed.) (2000) *The Letters of Charlotte Brontë: With a Selection of Letters by Family and Friends: Volume Two, 1848–1851*. Oxford: Clarendon Press.

Squire, S.J. (1994) The cultural values of literary tourism. *Annals of Tourism Research* 21 (1), 103–120.

Tetley, S. and Bramwell, B. (2002) Tourists and the cultural construction of Haworth's literary landscape. In M. Robinson and H.C. Andersen (eds) *Literature and Tourism* (pp. 155–170). London: Thomson Learning.

Thompson, C.J. and Arsel, Z. (2004) The Starbucks brandscape and consumers' (anticorporate) experiences of glocalization. *Journal of Consumer Research* 31 (3), 631–642.

Young, L. (2015) Literature, museums, and national identity; Or, why are there so many writers' house museums in Britain? *Museum History Journal* 8 (2), 229–246.

Zenker, S. and Braun, E. (2010) The place brand centre – A conceptual approach for the brand management of places. Paper presented at the 39th European Marketing Academy Conference (Copenhagen, 1–4 June).

3 *The Witcher* Novels and Games-inspired Tourism in Poland

Aleksandra Jaworowicz-Zimny

In 2015, shortly after its premiere, *The Witcher 3: Wild Hunt* became one of the gaming world's most acclaimed computer RPGs (role-playing game, where players create, develop and control a specific character who explores the game world engaging in a variety of quests). As a result, many fans turned their eyes towards Poland, the country in which the *Witcher* games were developed. The game itself is based on a series of novels by Polish author Andrzej Sapkowski, who created a narrative set in a fantasy world resembling medieval Eastern Europe which is deeply rooted in Slavic traditions, customs and mythology. The Polish origins of the franchise caused most dedicated fans to develop a broader interest in Poland, with some even being inspired to visit the country.

This chapter discusses tourism by *The Witcher* fans in Poland, with a particular focus on international participants of a three-day-long LARP (live action role-playing game) event, among whom a questionnaire was distributed. As *The Witcher* does not contain any particular real-life locations, touristic practices inspired by the franchise differ from pilgrimages by fans to sites specifically depicted in works of popular culture. Players express interest primarily in trips directly related to *The Witcher* narratives and game production, with general sightseeing in Poland in the areas resembling the game's visuals and atmosphere being of secondary value.

While numerous studies have been dedicated to film location tourism, including New Zealand tourism after the *Lord of the Rings* film trilogy (Buchmann, 2010; Carl *et al*., 2007) or the popularity of *Game of Thrones* tours in Northern Ireland (Tzanelli, 2016), as well as Harry Potter-inspired tours in the UK (Lee, 2012), video-game-induced tourism remains under-researched and major patterns remain unidentified. In his study of film-induced tourism, Roesch (2009: 231) pondered 'whether computer or video games can also trigger tourism in the future'. The answer currently seems to be positive. In Florence, guides offer *Assassin's Creed* tours (based on an action-adventure video games series set in different historical eras, with several installments in Renaissance Italy), introducing visitors to historical buildings featured in the games. Meanwhile, Sázava Monastery in the Czech Republic has reported an increase in the number of young visitors since the premiere of *Kingdom Come: Deliverance*, a historically accurate RPG set in medieval Bohemia (Oppelt, 2018). Videogame-induced tourism has also been discussed extensively in relation to Japan, where industry involvement in tourism promotion seems more established (Durango & Wei, 2016; Salmond & Salmond, 2016; Sugawa-Shimada, 2015; Yamamura, 2018).

While most tourists visiting Poland because of *The Witcher* have begun exploring the franchise by playing the game, they are familiar with the facts that the franchise exists across a variety of formats and that the main characters, narratives and locations have appeared in multiple media formats. Fans not only deepen their knowledge about the Witcher's World via media consumption, but actively extend its contents in their own productions like fan comics and fan fiction. Polish enthusiasts have even created a fan film, a non-profit crowd-funded full-length production set in the Witcher's World, but story-wise unrelated to video games and novels (WitcherFanFilm, 2019). Similarly, *The Witcher* LARP allows participants to create their own characters and storylines within the universe. *The Witcher* as contents not only exists in multiple forms, but also has become a living world with its own rules and characteristics, constantly expanded and recreated by fans. *Witcher*-induced tourism, therefore, is not simply game-induced tourism, but archetypal contents tourism.

The Witcher Phenomenon

Witcher Geralt of Rivia was first introduced to the public in 1986 in the short story *The Witcher* (*Wiedźmin* in Polish) by Andrzej Sapkowski published in the monthly magazine *Fantastyka*. Geralt is a professional monster slayer for hire. Magical mutations combined with heavy training give him supernatural abilities like enhanced sensory perception, and great speed, strength and endurance. He can use magic spells and elixirs that enhance his skills. Geralt travels through lands inhabited by different species including elves, humans and dwarfs. Equipped with two swords (silver to kill monsters and steel to kill humans) that are the trademark of his profession, the witcher is struggling to choose 'the lesser evil' in the times of social and political unrest.

The world of the witchers, sketched in the first short story, was soon developed by the author. The short story collections *The Witcher* (1990), *Sword of Destiny* (1992) and *The Last Wish* (1993) included altogether 14 stories centred on Geralt of Rivia. The massive success of the series in the Polish book market did not come, however, until the publication of *The Witcher* pentalogy (published between 1994 and 1999). Heavy in political content, the novels focus on Geralt and a witcher-in-training girl Ciri who had been raised by Geralt like a daughter. The novels have been translated into the majority of official European languages, as well as Chinese and Japanese.[1]

The massively popular series was soon adapted to other media forms. Polish adaptations include a comic book series (1993–1995), a live-action film (2001), a TV series (2002), a tabletop RPG game (2007), a card game (2007), three video games (2007–2015), a board game (2016) and a musical (2017). From the mid-2010s, *The Witcher* adaptations were no longer limited to Poland. Dark Horse has been publishing a comic books series since 2014. Moreover, in 2017 Netflix announced another TV series adaptation to premiere in 2019, raising the hopes of fans from all over the world that the franchise will get even more recognition.

The world created by Sapkowski gained worldwide popularity, particularly on the back of *The Witcher* video games. The first installment (*The Witcher*), developed by what was at the time a small studio called CD Projekt Red, was released in 2007. A sequel, *The Witcher 2: Assassins of Kings*, came out in 2011. Both games were RPGs for PC computers in which players took the role of Geralt and went on quests relating to events after the story of the pentalogy, not repeating, but extending the books' narrative. The first installment was set in a Poland-inspired landscape and deeply rooted in Polish classical literature and legends. The second game, with more complex gameplay and options for character development, was set in a more visually generic fantasy world. It allowed players to influence the storyline through the choices they made, thereby changing the in-game events and the game's final outcome. With this second game, CD Projekt Red built a sizable audience outside Europe. It even became a cultural icon in Poland to the point where in 2011, when US president Barack Obama visited the country, prime minister Donald Tusk gifted him a copy of *The Witcher 2* (Schreier, 2017).

After the international success of *The Witcher 2*, CD Projekt Red was ready for their biggest project yet. The next game was planned as a huge, open-world RPG, with the potential for 200 hours of gameplay (other popular open-world RPGs at the time were taking approximately 46 hours to finish) (Gera, 2015). Released simultaneously for PC and consoles (PlayStation 4 and Xbox) in May 2015, *The Witcher 3: Wild Hunt* sold over 4 million copies in two weeks (Crecente, 2015), receiving universal acclaim both from critics and gamers. During the 10 years since the original game was released, the series has sold over 33 million copies worldwide, with the majority of sales being of the third installment (Nelva, 2018).

The Witcher 3 as a Product of Polish Culture

The Witcher games series is a rare example of a Polish product that successfully captivated consumers all over the world. Not only was it developed in Poland, but numerous elements of the game world evoke associations with Poland. Sapkowski's novels have numerous references to Slavic myths and Polish classical literature (although these can be missed by readers unfamiliar with the topic), but rely on readers' imaginations in terms of storyline setting. The 'Polishness' of the Witcher's World is more obvious in its visual mediums, particularly *The Witcher 3*, which is also the adaptation that has received greatest international recognition. As such, the discussion will focus on the links of this game to Poland.

Konrad Tomaszkiewicz, one of *The Witcher 3* directors, stated in an interview for PlayStation Polska:

> We are from Poland and we take great pride in it. Our artistic team used Polish locations as references. Places included our picturesque fields, medieval castles and villages. Our artists photographed such places and tried to reproduce them in the game as faithfully as possible. *Witcher 3* is a Polish product which we export to the West. Foreigners do not know what Poland looks like, or Eastern Europe in general. In this sense, we offer something new and interesting. (PlayStation Polska, 2015)

As confirmed by Tomaszkiewicz, many visual aspects of *The Witcher 3* were modelled on the photographs taken by the developer's team members (a process also described by Yamamura, Chapter 4, and Butler, Chapter 5), resulting in elements of Polish landscapes, architectural features and symbols being present in the game.

The Land of Velen, one of the game's main locations, strongly resembles the rural Polish landscape, particularly the northern lake district of Masuria. Picturesque landscapes are completed with small villages, consisting of wooden huts with thatched roofs resembling medieval peasant houses. Some of the cottages are decorated with Polish folk patterns. Another main location, the free city of Novigrad, can be associated with the city of Gdańsk, Poland's main seaport. The former port crane and current symbol of Gdańsk (Figure 3.1) is present in Novigrad port (Figure 3.2) and

Figure 3.1 City of Gdańsk, Poland 2018. Author's Photo

Figure 3.2 Crane in the docks of Novigrad. Source: Screenshot from *The Witcher 3: Wild Hunt* © CD Projekt Red

will be easily recognized by anyone who has been to or seen pictures of Gdańsk. Furthermore, the national symbol of Redania, one of the Four Kingdoms in the Witcher's World, is a white eagle in a crown, holding a sceptre, placed against a red background. This is strikingly similar to the Polish coat of arms.

In addition to the visual layer, Polish (and more generally, Slavic) culture is represented in references to mythology, customs, legends and literature. Zaborowski (2015) has discussed the presence of Slavic mythology in *The Witcher* novels, but the creatures he mentions play much bigger roles in the games because killing monsters is the player's main task. In the game players encounter over 100 creatures, of which Slavic mythological creatures constitute around 90% according to game director Tomaszkiewicz (Culture.pl, 2016). Moreover, the game includes some recognizable Slavic pagan traditions and storylines referring to folklore and classic literature. Jakub Szamałek, the senior screenwriter, stated that the creators took multiple legends and folk tales but reworked them in a grimmer, cynical and less fairytale-like manner that would still be recognizable for the audience and simultaneously refreshed the source material in a manner befitting the dark world of *The Witcher* (Culture.pl, 2016).

The presence of Polish culture in *The Witcher* games is not only an element noticed by players with Slavic backgrounds, but also something discussed explicitly in different forums on the internet. From academic articles about the representation of Polish folklore and traditions in the games (Majkowski, 2018; Zaborowski, 2015), through to articles published on gaming portals (Miszczyk, 2015; Schreiber, 2017), to discussions on popular international forums (particularly reddit.com), the desire to explore *The Witcher*'s Polish roots is high among the players.

The Witcher and General Sightseeing in Poland

According to Social Surveys Department of Statistics Poland, in 2018 around 4,868,900 foreigners visited Poland for more than one day for the purpose of leisure, a 4.5% increase compared with 2017 (Główny Urząd Statystyczny, 2018: 101). With Lodz listed in second place in the Lonely Planet *Best Value Destinations 2019* (Lonely Planet, 2018) and with Wroclaw awarded the title of The European Best Destination 2018 (Best European Destination, 2018), Poland's significance on Europe's tourist map is slowly growing. However, in 2015 the top 12 most visited (by both domestic and foreign visitors) tourist attractions in Poland were five national parks, three museums located in royal castles and palaces, the Auschwitz–Birkenau Memorial, Wieliczka Salt Mine, Wrocław Zoo and the science museum in Warsaw (Polska Organizacja Turystyczna, 2016: 19). None of these is directly related to Polish folk and traditional culture.

The Witcher has the potential to introduce fans to numerous other places and events of interest in Poland, particularly those that are currently not recognized as the most popular destinations. Real locations reflected in the games, like Gdańsk or the Biskupin and Maurzyce open-air museums, would probably be of primary interest to dedicated gamers. Furthermore, Polish medieval culture was massive inspiration for *The Witcher*. Medieval castles and numerous knights' tournaments and battle re-enactments organized in historical places could, if advertised, attract visitors fascinated with the Witcher's World. Finally, places and events promoting Polish folklore and traditional tales like the Ethnographic Museum of Kraków and Kupala Night[2] celebrations that occur in various places in Poland in the second half of June are deeply related to *The Witcher* narratives and characters. The franchise, therefore, has the potential to trigger wider touristic interest in Polish medieval and folk culture.

Browsing *The Witcher* section of Reddit.com and the CD Projekt Red forum reveals several threads by people seeking advice about sites worth visiting in Poland. Prospective visitors look for recommendations of historical, particularly medieval, locations that resemble the Witcher's World, but also thematic pubs, events and stores (cf. Deeshon, 2017; shartweekondvd, 2017). Many fans are teenagers or in their early twenties, and so have financial limitations in undertaking such trips. On the blog of Plan Poland – a small company operating mainly online that organizes customized trips to Poland according to customers' budgets, time limits and interests – the most popular post is 'Polish culture and real places in "The Witcher 3: Wild Hunt"' (Plan Poland, 2017). This blog entry presents the literature, customs and real-life locations that inspired specific aspects of the game. Justyna Dzik, a tour organizer from Plan Poland and a *Witcher* fan herself, told me that she was hoping for major interest in *The Witcher*-inspired tours. The company has promoted such tours on forums and social media where fans of the franchise gather, but although several people expressed an interest in visiting Poland at some point, there was no major response to the offer, and no single request for such a tour to be organized. While gamers do not seem overly interested in sightseeing, Ms Dzik hopes that fans of the upcoming Netflix *Witcher* TV series will be, as the show will be shot partially in Poland (Dzik, 2018).

Private initiatives seem insufficient to induce broader *The Witcher* tourism, even though fans seem interested in events, attractions and places related to their favourite franchise. Place branding is one of the elements influencing consumers' associations with a location (see Thyne & Larsen, Chapter 2), but as of October 2018 no strategies using *The Witcher* to promote locations and events in Poland had been implemented on either local- or state-level government levels. The franchise's connection to Poland's medieval heritage and folklore has the potential to influence tourism, but without purposeful, organized branding remains unrecognized.

CD Projekt Red Offices – Trips for the Lucky Ones

Multiple fans ask on the CD Projekt Red internet forum for a very specific type of trip: a visit to the Warsaw-based offices of *The Witcher* developers. People who, for various reasons, are travelling to Poland express interest in learning about where the game was made, meeting the developers and buying goods in the official *Witcher* merchandise store they expect to be located by the game studio. However, to the major disappointment of many fans, there is no Witcher-dedicated store in Poland other than an online official one run by CD Projekt Red. On thematic forums questions about the office's location, open days, tours or how to contact the developers can be found (cf. Br0adsw0rd, 2010; Vitalyaya, 2016).

Even though fans would be interested in such a tour, it is currently not available for all enthusiasts. Before the release of *The Witcher 3*, when the studio was less recognized, entering it seemed less challenging. In 2014, a fan posted pictures and a short report from his trip to the offices, which he was allowed to enter simply by emailing the developers a few days in advance (Cyph0n, 2014). However, with the massive growth of the studio after *The Witcher 3*, visiting it requires a more established position in the gaming community, leaving the option available only for bloggers, YouTubers and writers from the popular gaming portals.

According to Radek Grabowski, PR senior manager, as of 2018 the only visits allowed are those by community representatives, which includes 'Thank you' events for the official *The Witcher* forum users (Grabowski, 2018). Most active users are invited to the studio on such occasions, at which time they can look around freely, see materials about the making of the games, and ask questions to the CD Projekt Red staff (CD Projekt Red, 2017). While events and promotional tours for celebrities of the gaming world target the international community and are held in English, forum users reunions focus on the Polish-speaking community within the country.

None of the forum users stated they would like to go to Poland especially to visit the developer's studio. Most mentioned it as an extra activity during their stay in Poland on a different occasion. However, it is a place many fans would be willing to add to their itineraries. The potential of the developers–gamers relationship can be seen in the example of CyberConnect2, a Japanese game development studio. CyberConnect2 organizes thematic exhibitions, seminars and meetings with the company's president for fans, as well as offering company tours for anyone who registers in advance (CyberConnect2, n.d.). Visiting the offices is of special value for anyone interested in a potential career in the industry, but also for fans who can receive first-hand information about their favourite franchise. It can be seen as similar to the off-location tourism to film sets that 'provide opportunities for industrial-style tourism activities such as tours of the working studios and hands-on experiences with the technology' (Beeton, 2005: 28). *Witcher* fans cannot visit actual locations that would fully correspond with the animated reality they observed in the game, but they are attracted to authentic sites related to its creation and the closed world of games production.

'The Witcher School' – Attracting Fans from All Over the World

In 2015, a *Witcher*-related event bringing fans from all over the world to Poland was introduced. The Witcher School (hereafter TWS) is a three-day-long LARP organized by the group 5 Żywiołów in the Silesia region of Poland.

LARP has its roots in tabletop role-playing games in which 'one or more players take on roles within an interactive story, usually under the guidance of one or more game masters' (Tychsen *et al.*, 2006: 252–253). In this practice, which was developed in the 1970s, participants describe their characters and actions during the gaming session. LARP, however, is 'a roleplaying game where participants are required to inhabit their character/role with their physical bodies, and where interaction with a material setting defines the game' (Regitzesdatter, 2011: 74). In this case, actions are physically performed by the participants, even if different levels of abstraction and simulation are maintained in the game (Stenros & Hakkarainen, 2003: 59–61).

The Witcher School takes place in one of two locations in southeast Poland, the 12th-century Grodziec Castle or the 17th-century Moszna Castle. During the event all modern technologies in the location are concealed to help the players to get into their role. Upon arrival players receive their costumes and attend workshops explaining safety rules and the basics of the game. Many of them are primarily *Witcher* fans, and LARP is a new experience for them. Participants receive descriptions of their characters and the basic scenario in advance. Some minor characters known from the franchise are included into the LARP scenario, but the storylines are unrelated to the books and games. Consequently, *The Witcher* contents are actively recreated and expanded by LARP participants. Players embody original witcher apprentice characters in the realms of the Witcher's World, using elements of Polish landscape and architecture resembling those known from the fictional narratives. They participate in fencing training, archery, meditations or alchemy workshops. Sometimes they are woken up in the middle of the night, required to fight monsters, run through the cold forest and raid the castle.

The first Witcher School was conducted in Polish in 2015, but the following year, in March 2016, the first International Witcher School was held in English. It attracted 39 players (Purchese, 2016). In 2018, there were six editions planned, three in Polish and three in English, and all six events were fully booked. Depending on the location, the number of participants varies from 65 (Grodziec Castle) to 128 (Moszna Castle). On 22 October 2018, ticket sales for three Spring 2019 events began. One week later, on 29 October, tickets for the Polish edition (costing €310, with 65 tickets available) and for the April English Edition (costing €430, with 128 available) were sold out, and for the March English edition (costing €430, 128 tickets available) there were only 20 left. Even though TWS is promoted only on the internet, information exchange on Facebook and former participants' blog entries on popular gaming portals (like Eurogamer.net) are enough to attract growing attention to the event.

Survey Among Former TWS Participants

One of TWS's characteristics is the strong social bonding that occurs during the event. Ex-witchers build a close community and remain in touch both in real life and through social media. A closed Facebook group moderated by 5 Żywiołów connects over 200 former participants. Although I did not participate in the event, with the cooperation and permission of the LARP organizers, a questionnaire asking about participants' experiences of travelling to Poland was posted. Eighty-six participants from 27 countries submitted answers.

The informants, 31 females and 55 males, are mainly Europeans, but also Americans, Asians and African (Figure 3.3). The majority of the informants are in their twenties, with the youngest being 19 and oldest 46. For the vast majority of them (72 out of 86), participating in TWS was their first trip to Poland, and 84 travelled especially with the purpose of participating in the LARP.

The majority of the participants decided to participate in the event mainly because of their great enthusiasm for the franchise, although some of them are also fans of LARP in general. Around 30% of the informants were motivated primarily by their interest in LARP. Even though the games are by far the most popular form in which *The Witcher* has conquered the world, TWS participants are familiar with the franchise in its

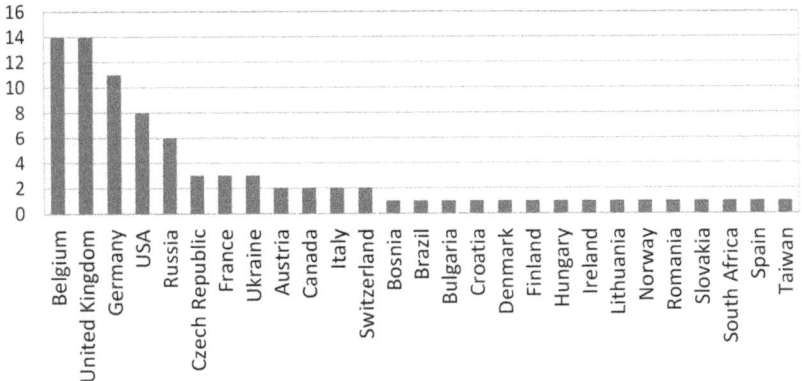

Figure 3.3 Nationalities of TWS participants

various forms, suggesting that dedicated fans tend to consume fictional narratives in all available ways (Figure 3.4).

One of the purposes of the survey was to establish the connections fans may see between the fantasy land of *The Witcher* and Polish landscape and culture. Moreover, there was a question regarding whether visiting Poland influenced this association in any way. Eighty participants answered the question about seeing a resemblance between the Witcher's World prior to visiting Poland, and 73 answered the question about their impressions after visiting Poland. Before the visit 39 participants (49% of the answers) did not see any resemblance between the worlds or did not think of it. After coming to Poland, the number was reduced to 17 answers (25%). Visiting Poland, therefore, caused some of the fans to notice the Polish roots of *The Witcher* they had not seen earlier.

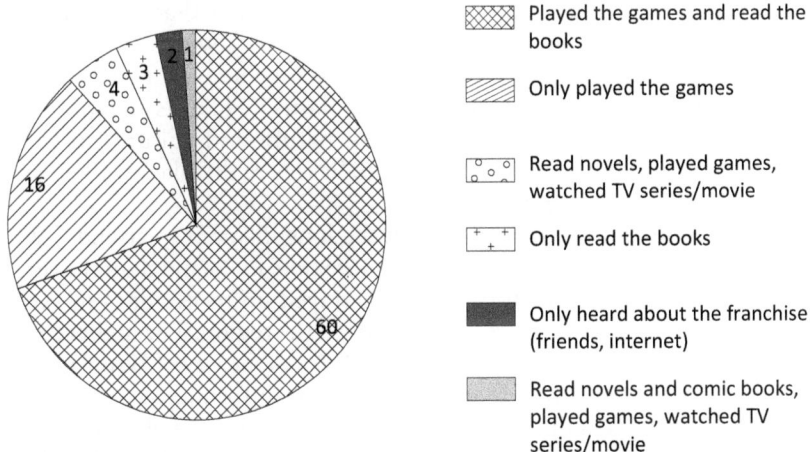

Figure 3.4 TWS participants' familiarity with *The Witcher* franchise

Among the participants who associated *The Witcher* with Poland even before visiting the country, the most commonly mentioned elements were Polish (and Slavic) mythology and folk tales. Other informants saw similarities between Polish landscape (eight answers) and architecture (seven answers), particularly medieval castles and small villages. Participants also recognized outfits, characters' designs and the heraldic symbols of Redania as Polish. One person even saw the political conflict in *The Witcher* as a metaphor for World War II history and Nazi Germany's invasion of Poland. Few informants mentioned researching the background of the game, although when this background was mentioned the numerous internet pages connecting *The Witcher* to Polish culture turned out to be very informative and had introduced Polish folklore to them. Despite CD Projekt Red's massive inspiration from Polish scenery, *The Witcher*'s landscapes were not automatically associated by most players with Poland. Elements of the narrative, however, are broadly discussed among fans and recognized as belonging to Slavic folklore.

Some visitors remain sceptical regarding the Witcher's World's connection to Poland. Some informants explained that they cannot see similarities between the medieval-like reality of *The Witcher* and 21st century Poland, or that it is Slavic culture rather than Polish, or that they do not like the idea that the world of a RPG could in a way 'belong' to a specific country because a RPG is place 'where everyone can find something that reminds him perhaps of home and where everybody can feel like home, regardless of his origin [sic.]'. Nevertheless, participation in TWS resulted in 19 informants noticing the Polish roots of *The Witcher* that they had not seen before, and three people saw the correlation 'a little'. Elements that caught players' attention were mainly landscapes and medieval architecture, particularly Moszna and Grodziec castles. Although one person noted that the similarities are visible mainly in the rural areas and harder to spot in big, modern cities, some claimed that travelling to Gdańsk (the model for Novigrad) and Krakow (which has a well-preserved medieval centre) evoked the atmosphere of *The Witcher*. Overall, after visiting Poland, 61 informants (71%) associated *The Witcher* with different aspects of Polish nature and culture[3] to some point, with the spectrum from 'a little' to 'it was really like I walked and trained … IN the Witcher world!'

Since many of the participants travel a long distance to attend TWS, some of them use the occasion to spend more time in Poland than the three days of the main event. More than half stayed up to five days in total, and used the extra two days for sightseeing in the region. The main attraction for the participants other than the LARP was the city of Wroclaw, the capital of the region, which was visited by 34 informants. Sightseeing tours in the city, including visits to the famous city zoo and old town, were a popular way for participants to spend some extra days in Poland. Others who had more time visited both very famous tourists sites, like Warsaw,

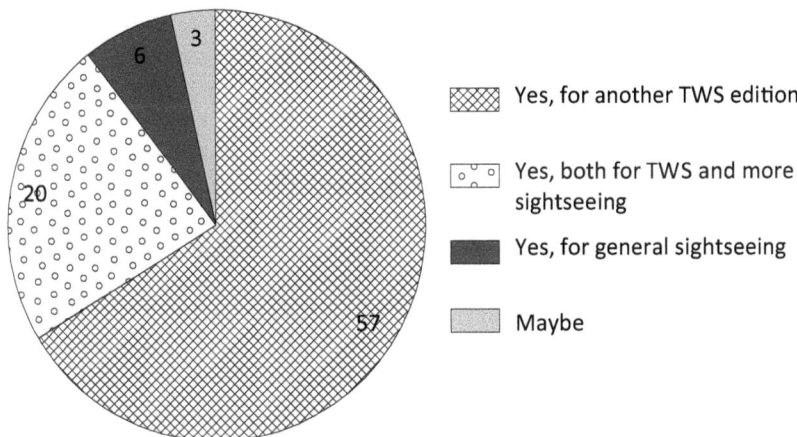

Figure 3.5 TWS participants' plans for visiting Poland again

Krakow, Auschwitz and Gdańsk, as well as some less popular sites, including Poznan, Nysa and Prudnik. These TWS participants were interested in returning to Poland, and some considered coming back for a general sightseeing trip (Figure 3.5).

Finally, TWS is an adventure that strongly bonds its participants and influences their future travel plans. Several informants mentioned visiting friends or meeting new people as reasons to come back to Poland in the future. Moreover, former participants organize 'Witcher meet ups'. They invite fellow 'witchers' to their hometowns in different countries for sightseeing and enjoying each other's company. This bottom-up initiative was not suggested in any way by LARP organizers. Trips' destinations are not related to *The Witcher* in any way, but still involve narrative-motivated travel. Unlike fan conventions, such meetings do not include organized events or commercial activity. They allow like-minded enthusiasts who bonded in the Witcher's World to relive their experience, discuss their fandom and foster their new way of life resulting from TWS participation. Numerous players have stated in comments on the TWS Facebook page that they quit smoking, started working out and lost weight with the purpose of becoming better witchers during the next LARP event. The 'human factor' and fan networks are an important travel motivation. They become a natural extension of the places and events more directly related to the fictional contents.

Conclusions

The Witcher in all available media forms presents a narrative deeply rooted in elements of Polish culture, particularly folk tales. For the international fan community the most common point of entry to the franchise were the video games, which use visuals based on the heraldry, architecture and landscape of Poland. A great amount of narrative content is also based on Slavic folklore and introduces consumers to monsters and storylines recognizable from regional myths. Despite being a fantasy realm, the Witcher's World is recognized by many fans as Poland-based, which results in numerous internet posts engaging in in-depth analysis and the search for possible similarities. Moreover, *The Witcher* is recognized as a product made in Poland, evoking interest in game development in the region.

Even though many fans are aware of (particularly narrative) similarities between Poland and the Witcher's World, this fact in itself is not necessarily enough to induce travel to Poland, particularly considering the limited budget of young fans. In their online discussions, fans express interest in *Witcher*-related attractions, events, goods and locations that would allow them to feel closer to the narrative they admire. Despite the potential of Poland's medieval heritage, which could attract *Witcher* fans, such places lack systematic promotion targeting the group. As a result, many fans remain unaware of places that could be considered extensions of the *Witcher* narrative, and consequently have little motivation to undertake a trip to Poland.

As of October 2018, the event most clearly attracting *The Witcher* fans from all over the world to Poland remains the live-action roleplaying game, TWS. The event enables physical adventure in the Witcher's World, in a narrative consistent with what fans of the franchise grew to know and admire. Players get to reconstruct and expand the narrative in a physical environment strikingly similar to what they saw in visual representations of *The Witcher*, particularly the video games. Rather than passive consumption of a game-related place or recreation of a narrative known from the franchise, LARP players engage in active creation of their own stories within *The Witcher* setting. Much like *The Witcher* video game allows gamers to shape the plot, the immersive experience of TWS allows players to influence the storyline, making players not only consumers, but also participants in the narrative. This is enthusiastically received by attendees.

The Witcher School participants get a chance to discover Poland primarily in its less touristic, rural version that they can recognize as landscapes known from *The Witcher* games. Moreover, many participants start to associate elements of Polish culture, like customs, legends and language, with the Witcher's World. As a result of the positive TWS experience, the participants are likely to develop broader interest in sightseeing tours in Poland, visiting again not only for another LARP session. The visibly growing popularity of TWS suggests the touristic potential of *The Witcher*-related events in Poland, with the prospect of eventually increasing general tourism in the country.

Finally, one of the crucial elements of TWS experience is the bonding of players. After shared adventures in the Witcher's World, they keep closely in touch. They not only sign up for upcoming editions of TWS together, but also invite each other for group meetings in their hometowns in different parts of the world. In this way *The Witcher* 'contents tourism' is not limited to Poland, but can occur in any place where fans will travel with the purpose of sharing and celebrating their devotion to the franchise.

Notes

(1) As of October 2018, short stories collections and four volumes of the pentalogy were translated into Chinese, and the first three volumes of the pentalogy were translated into Japanese.
(2) Slavic ritual related to the summer solstice. Today celebrations include bonfires, traditional dances, floating wreaths and traditional craft shows and workshops.
(3) In this question, 54 people confirmed seeing the resemblance after visiting Poland. Seven of the informants who stated that they saw similarities even before the visit left the question about 'after' unanswered. Since these informants did not give any explanation about not seeing similarities anymore after the trip, they were treated and counted as those who see the similarity on the assumption that their opinion remained unchanged.

References

Beeton, S. (2005) *Film-Induced Tourism*. Clevedon: Channel View Publications.
Best European Destination (2018) Best places to travel in Europe 2018. See https://www.europeanbestdestinations.com/european-best-destinations-2018/ (accessed October 2018).
Br0adsw0rd (2010) Visiting Poland. See https://forums.cdprojektred.com/forum/en/the-witcher-series/general-discussions/24847-visiting-poland (accessed May 2018).
Buchmann, A. (2010) Planning and development in film tourism: Insights into the experience of *Lord of the Rings* film guides. *Tourism and Hospitality Planning & Development* 7 (1), 77–84.
Carl, D., Kindon, S. and Smith K. (2007) Tourists' experiences of film locations: New Zealand as 'Middle-Earth'. *Tourism Geographies: An International Journal of Tourism Space, Place and Environment* 9 (1), 49–63.
CD Projekt Red (2017) 10 lat serii gier o Wiedźminie – dziękujemy za spotkanie! See https://pl.cdprojektred.com/news/10-lat-serii-gier-o-wiedzminie-dziekujemy-za-spotkanie/ (accessed May 2018).
Crecente, B. (2015) The Witcher 3 sells 4 million copies in two weeks. See https://www.polygon.com/2015/6/9/8751113/the-witcher-3-sales (accessed May 2018).
Culture.pl (2016) Behind the scenes: The Witcher's Polish inspirations. See https://www.youtube.com/watch?v=xdhQnMNZqDw (accessed May 2018).
CyberConnect2 (n.d.) Kaisha kengaku. See http://www.cc2.co.jp/company/visit (accessed September 2018).
Cyph0n (2014) I visited CDP Red offices in Warsaw today! See https://www.reddit.com/r/witcher/comments/2dc9ey/i_visited_cdp_red_offices_in_warsaw_today/ (accessed May 2018).

Deeshon (2017) Big Witcher fan, what should I go looking for in Poland this summer? See https://www.reddit.com/r/witcher/comments/6a7u89/big_witcher_fan_what_should_i_go_looking_for_in/ (accessed May 2018).

Durango, B. and Wei, H. (2016) Games as travel guides: A look at meaningful downloadable content and its connection to locational context. *DiGRA Conference Proceedings* 13 (4), 1–9.

Dzik, J. (2018) Interview concerning prospects of The Witcher tourism (personal communication, April).

Gera, E. (2015) The Witcher 3 can take over 200 hours. See https://www.rockpapershotgun.com/2015/04/04/the-witcher-3-length/ (accessed May 2018).

Główny Urząd Statystyczny (2018) *Turystyka w 2017r*. Warszawa: Zakład Wydawnictw Statystycznych.

Grabowski, R. (2018) Interview concerning The Witcher tourism in the CD Projekt Red offices (personal communication, April).

Lee, C. (2012) 'Have magic, will travel': Tourism and Harry Potter's United (Magical) Kingdom. *Tourist Studies* 12 (1), 52–69.

Lonely Planet (2018) Best in travel 2019. Best value. See: https://www.lonelyplanet.com/best-in-travel/value (accessed October 2018).

Majkowski, T.Z. (2018) Geralt of Poland: The Witcher 3 between epistemic disobedience and imperial nostalgia. *Open Library of Humanities* 4 (1). See https://olh.openlibhums.org/articles/10.16995/olh.216/ (accessed May 2018).

Miszczyk, M. (2015) World games, diversity, outrage, and The Witcher 3. See http://nichegamer.com/2015/06/07/world-games-diversity-outrage-and-the-witcher-3/ (accessed May 2018).

Nelva, G. (2018) The Witcher series' sales pass 33 million units; PS4 vs Xbox One vs PC sales for the Witcher revealed. See https://www.dualshockers.com/witcher-sales-ps4-xbox-one-pc/ (accessed May 2018).

Oppelt, R. (2018) Herní hit Kingdom Come nastartoval v Sázavském klášteře turistiku. See https://cestovani.idnes.cz/sazava-klaster-kingdom-come-cirkevni-restituce-rataje-nad-sazavou-10a-/po-cesku.aspx?c=A180326_131116_po-cesku_hig (accessed May 2018).

Plan Poland (2017) Polish culture and real places in 'The Witcher 3: Wild Hunt'. See http://planpoland.com/realplacesinthewitcher/ (accessed May 2018).

PlayStation Polska (2015) The Witcher 3 on PS4 | Behind the scenes. See https://www.youtube.com/watch?time_continue=231&v=di3Zzp5yR_0 (accessed May 2018).

Polska Organizacja Turystyczna (2016) *Frekwencja w atrakcjach turystycznych w latach 2011–2015*. Kraków-Warszawa: Polska Organizacja Turystyczna.

Purchese, R. (2016) My gruelling weekend at Witcher School. See https://www.eurogamer.net/articles/2016-04-08-my-gruelling-weekend-at-witcher-school (accessed May 2018).

Regitzesdatter, J. (2011) Doing gender at LARP. In T.D. Henriksen, C. Bierlich, H.K. Friis and V. Kølle (eds) *Think LARP. Academic Writings from KP2011* (pp. 70–85). Copenhagen: Rollespilsakademiet.

Roesch, S. (2009) *The Experiences of Film Location Tourists*. Bristol: Channel View Publication.

Salmond, M. and Salmond, J. (2016) The gamer as tourist. The simulated environments and impossible geographies of videogames. In P. Long and N.D. Morpeth (eds) *Tourism and the Creative Industries: Theories, Policies and Practice* (pp. 151–163). Abingdon: Routledge.

Schreiber, P. (2017) How the Witcher plays with Polish romanticism. See http://culture.pl/en/article/how-the-witcher-plays-with-polish-romanticism (accessed May 2018).

Schreier, J. (2017) *Blood, Sweat, and Pixels: The Triumphant, Turbulent Stories Behind How Video Games Are Made*. [e-Reader version]. See https://play.google.com/store/books/details/Blood_Sweat_and_Pixels_The_Triumphant_Turbulent_St?id=-bK-DQAAQBAJ&hl=en_US (accessed May 2018).

shartweekondvd (2017) Witcher events/sights in Poland? See https://www.reddit.com/r/poland/comments/7cfcay/witcher_eventssights_in_poland/ (accessed May 2018).

Stenros, J. and Hakkarainen, H. (2003) The Meilahti School: Thoughts on role-playing. In M. Gade, L. Thorup and M. Sander (eds) *When Larp Grows Up: Theory and Methods in Larp* (pp. 54–65). Copanhagen: Projektgruppen KP03.

Sugawa-Shimada, A. (2015) Rekijo, pilgrimage and 'pop-spiritualism': Pop-culture-induced heritage tourism of/for young women. *Japan Forum* 27 (1), 37–58.

Tychsen, A., Hitchens, M., Brolund, T. and Kavakli, M. (2006) Live action role-playing games: Control, communication, storytelling, and MMORPG similarities. *Games and Culture* 1 (3), 252–275.

Tzanelli, R. (2016) Game of Thrones to game of sites/sights. Framing events through cinematic transformations in Northern Ireland. In K. Hannam, M. Mostafanezhad and J. Rickly (eds) *Event Mobilities: Politics, Place and Performance* (pp. 52–67). Abingdon: Routledge.

Vitalyaya (2016) Visiting CDPR's studio in Warsaw (Question). See https://www.reddit.com/r/witcher/comments/4u4gwp/visiting_cdprs_studio_in_warsaw_question/ (accessed May 2018).

WitcherFanFilm (2019) Pół Wieku Poezji Później – Alzur's Legacy. See https://www.facebook.com/witcherfanfilm/ (accessed April 2019)

Yamamura, T. (2018) Pop culture contents and historical heritage: The case of heritage revitalization through 'contents tourism' in Shiroishi city. *Contemporary Japan* 30 (2), 144–163.

Zaborowski, P. (2015) Mitologia słowiańska w cyklu o wiedźminie. In R. Dudziński, A. Flamma, K. Kowalczyk and J. Płoszaj (eds) *Wiedźmin. Bohater masowej wyobraźni* (pp. 21–33). Wrocław: Stowarzyszenie Badaczy Popkultury i Edukacji Popkulturowej 'Trickster'.

4 Travelling *Heidi*: International Contents Tourism Induced by Japanese Anime

Takayoshi Yamamura

There was a major turning point in postwar Japan in the 1970s regarding the 'multi-use of contents' based on creators' tourism experiences. It was the creation of a series of TV anime (cartoons) based on Western children's literature, beginning with *Arupusu no Shōjo Haiji*, or *Heidi, Girl of the Alps* (unless otherwise indicated, hereafter *Heidi* refers to this anime series), which was first broadcast in Japan in 1974 (Figure 4.1). In this chapter, I focus on this turning point, specifically the creative process behind *Heidi*. It clarifies how, through their own experiences of transnational contents tourism, the anime creators reinterpreted, recreated and 'remediatized' the narrative world. Recounting this story here demonstrates part of the dynamic process of 'contentsization'.

As I mentioned in the Introduction to this book, one of the major characteristics of contents tourism studies is a focus on 'the continual process of the development and expansion of the "narrative world" through both mediatized adaptation and tourism practice', namely 'contentsization' and 'the relationship between the multi-use of contents and tourist phenomena/practices by considering tourist sites as media'. As fans of an original work (whether *Heidi* or other source materials) and as contents tourists themselves, creators play a vital and central role in the process of contentsization because they reinterpret, re-edit and mediatize a given narrative world. It can also be said that mass contents tourists are their followers. *Heidi* is a typical example in which we can observe the continual process of the reinterpretation, re-editing and recreation of a narrative world, from the viewpoint of cross-referencing among media, particularly the original version, other works and physical sites as media.

There are three main reasons why Heidi is a particularly revealing example. First, *Heidi* is a typical example of the transnationalization and

Figure 4.1 The opening scene of the anime *Arupusu no Shōjo Haiji* (*Heidi, Girl of the Alps*), 1974. © Studio 100 Media Kindly supported by Studio 100 Media

transmediatization of the contents/'narrative world' with a long history. The Heidi story was originally written as a children's book in 1880–1881 by the Swiss storywriter Johanna Spyri (1827–1901). The work became popular in Japan and was adapted for use in various media. For example, since the Japanese translation of Spyri's novel was first published in 1920, there have been over 30 Japanese translations (Wissmer, 2015: 2) and a musical version was performed by the Takarazuka Revue (an all-female theatre troupe) in 1940.[1] In Western countries, including the US, Switzerland, Germany, Austria and the UK, the Heidi story has been adapted as live-action films. To the author's knowledge, the 1974 Japanese anime *Heidi*, produced by the animation studio Zuiyo Eizo, was the first anime version of the Heidi story in the world. Therefore, focusing on *Heidi* allows us to understand the process of the transnational/transmedia reinterpretation and reconstruction of 'Heidi's narrative world'. In particular, *Heidi* is significant as a case of Japanese anime creators (non-Western, non-literary creators) reinterpreting Swiss (Western) literature.

Moreover, after the broadcast of *Heidi* in Japan in 1974, it was broadcast outside Japan, first in Spain and then in Italy and West Germany in the 1970s. *Heidi* was a big hit in these countries, according to Tomosuke Suzuki, then Japan's representative and sales agent for EMTV Studio 100 (interview by the author, 27 September 2018).[2] Thereafter, it was broadcast in many other countries, and now the anime image of Heidi is internationally recognized.

The second reason is that *Heidi* was the first TV anime series in Japan for which the creators went practical location-hunting abroad (Chiba, 2017: 5). In other words, they travelled abroad as literary or contents tourists and their transnational tourist experiences can be observed clearly in the production process. The creators, who went on to become some of the most famous names in the Japanese anime industry, went on a location-hunting tour for *Heidi*. Their experiences greatly impacted the anime creation process in Japan, particularly in the establishment of a 'layout system'[3] and in the search for reality in fiction[4] (Hikawa, 2017: 77). The tour members were Isao Takahata (director), Hayao Miyazaki (layout), Yoichi Kotabe (character design, animation director), Takeo Watanabe (music), Junzo Nakajima (producer) and Shigeto Takahashi (executive producer). This in-depth location hunting was a turning point. It became customary thereafter for the subsequent anime series in World Masterpiece Theater (*Sekai meisaku gekijō*, 1975–1997, 2007–2009) based on Western children's literature to go location hunting (Takahata, 2008: 94). Moreover, the culture of location hunting for anime films has been adopted by other anime filmmakers, such as Ghibli, in which Takahata and Miyazaki became central figures. Many other anime works made using this location-hunting culture continue to generate a huge following of contents tourists (Okamoto, 2014, 2015; Yamamura, 2015).

Third, *Heidi* stimulated Japanese tourism to Switzerland and the image and narrative world (or tourism imaginary) reconstructed and developed by anime has been broadly and internationally accepted up to the present day. For example, many local tourism companies and facilities continue to use images originally based on the Japanese anime (Figures 4.2 and 4.3). Moreover, although Josef the St Bernard, who was created for the anime version, does not feature in the original book (Bandai Visual, 2011: 6; Wissmer, 2015: 161, 211), his image is now widely used for local tourism promotion and attractions.

The research for this chapter is based mainly on contemporary records, including publications by and interviews with the creators.[5] I also interviewed Mr Junzo Nakajima, the producer of *Heidi*, Ms Kaori Chiba, the *Heidi* and children's literature/anime researcher, and Mr Tomosuke Suzuki, then Japan's representative of Beta-Film in 1970s. In addition, in July 2014 and August 2018 I conducted fieldwork and materials collection at tourist sites related to *Heidi* and at the sites discovered during the original location hunting in 1973 in Switzerland. I also communicated personally by email to people or related organizations such as Heididorf. In this manner, the case study is based on qualitative analysis of original documents, media, websites and interviews, and uses the 'triangulation' of sources (Patton, 2001).

Figure 4.2 A mixed-media image of Heidi's narrative world. This is the official poster image of Heididorf, Maienfeld, Switzerland. The anime images of Heidi (left) and Peter (right) can be seen with the image of Johanna Spyri (top right) and other media images. The company Heididorf runs Heididorf ('Heidi village') in Maienfeld for international tourists. It was founded in 1998[6]

Source: Image courtesy of Heididorf

Figure 4.3 The anime image of Heidi in front of the entrance to the Johanna Spyri museum, Hirzel, Switzerland (August 2018). Author's photo

The Heidi Location hunting Tour: Accessing and Re-interpreting the Original Narrative World via Real Sites

A detailed itinerary of the two-week tour is given in Table 4.1. The year 1973, when the tour was conducted, was just nine years after the 1964 liberalization of Japanese foreign travel by the Japanese government. The tour was the first location-hunting tour abroad in Japanese TV anime history.[7]

The contemporaneous records, including interviews, illustrate that the tour group engaged in intensive and in-depth research, not only by sketching the landscapes and people, but also by researching the socio-cultural attributes of the place as a whole. They worked hard to gain access to and experience the original narrative world of Heidi in Switzerland and Germany (Figure 4.4). Their comments also reveal that they went location hunting to search for the reality in the fantasy. Hayao Miyazaki talked of recreating 'a respectable narrative world' (Bandai Visual, 2010: 6) and of how director Isao Takahata searched for realism using anime as the medium so that 'people won't feel something is strange, even if local Swiss people watch it'. Takahata had 'watched many Western films that depict Japan from a basis of misunderstanding, such as a character walking in a tatami room with wooden clogs' (Miyazaki, 2005). As the following discussion makes clear, the tour changed their vision of the imaginary world

Table 4.1 Heidi location-hunting tour in July 1973

List of participants

- Isao Takahata (director)
- Hayao Miyazaki (scene design, layout, screenplay)
- Yoichi Kotabe (character design, animation director)
- Junzo Nakajima (producer)

  This itinerary pertains only to these four members. The following two members left for Switzerland earlier and were later joined by the four participants above.

- Takeo Watanabe (music)
- Shigeto Takahashi (president of Zuiyo enterprise company, exective producer)

Schedule		Stage
15 July	Tokyo Haneda → (flight) → Anchorage → (flight) →	Flying overnight
16 July	→ (flight) → Amsterdam, Schiphol → (flight) → Paris-Orly → Zurich	Zurich
17 July	Location hunting day 1: Zurich	Zurich
18 July	Location hunting day 2: Zurich → Chur → Maienfeld	Maienfeld
19 July	Location hunting day 3: Maienfeld, visited Heidi's Alm hut	Maienfeld
20 July	Location hunting day 4: Maienfeld, visited 'Dörfli' and 'Heidihof',	Maienfeld
21 July	Location hunting day 5: Takahata, Kotabe, Miyazaki took a day trip to St Moritz	Maienfeld
22 July	Location hunting day 6: Maienfeld → Chur → Spiez → Interlaken → Grindelwald → Kleine Scheidegg	Kleine Scheidegg
23 July	Location hunting day 7: Kleine Scheidegg → Jungfraujoch → Lauterbrunnen → Interlaken	Interlaken
24 July	Location hunting day 8: Interlaken → Bern → Lucerne → Zurich → Frankfurt	Frankfurt
25 July	Location hunting day 9: Frankfurt	Frankfurt
26 July	Location hunting day 10: Frankfurt	Frankfurt
27 July	Frankfurt → (flight) →	Flying overnight
28 July	→ (flight) → Tokyo Haneda	

Sources: Prepared by the author based on Kotabe (2013: 81–83), Chiba (2008: 77) and Chiba (2017: 72–84). Supplemented by information in interviews with Junzo Nakajima and Kaori Chiba.

Figure 4.4 A group photo in front of a hut in Ochsenberg, Maienfeld, taken by Junzo Nakajima during the location hunting in July 1973 (from left, Hayao Miyazaki, Yoichi Kotabe, and Isao Takahata)
Source: Photograph courtesy of Junzo Nakajima.

of Heidi and they reinterpreted Heidi's narrative world based on their communication with local people and through site visits. They attempted to access and enhance the sense of reality of the narrative world of Heidi through their tour and they cross-referenced the real sites with the original narrative world. Their experiences as contents tourists, therefore, made it possible for them to reinterpret and remediatize the narrative world of Heidi into a 'respectable' and high-quality anime version.

All of the tour participants of had engaged in extensive research before the tour, and this then informed their experiences while on the tour. Director Isao Takahata recalled:

> I read Heidi when I was a child and I loved it. [...] We were very impressed by the nature in Switzerland and we conducted an intensive survey with our eyes and hearts without prejudice. This experience helped with the production of the anime *Heidi*. Moreover, we bought two versions of the Heidi book in Zurich with old illustrations. And the picture books by Arois Carigiet, which had already been published in Japan, helped us greatly with the description of daily life. [...] We used Goethe Haus in Frankfurt as a reference for Sesemann's house. Undoubtedly, the reason why we depicted the world of Heidi as far as possible based on on-site research was because we took it for granted that this should be done, not because we wanted to show our work to foreign people. [...] We tried to create *Heidi* for the Japanese audience. (Takahata, 2008: 90, 94–95).

Commenting on these sentiments, Kotabe noted, 'Takahata believed that we could depict their daily life in detail because the *Heidi* anime was a one-year TV series (52 stories). So, it helped a lot that we went location hunting and could see the related sites directly' (Kotabe, 2013: 90).

One of the key sites visited was the Alm hut. Takahata described how 'after we had seen the museum in Zurich, we stayed in Maienfeld for about five days' (Takahata, 2008: 94).

> Dörfli village in the original Heidi story is fictional; however, in Maienfeld, we visited an abandoned building modelled on the winter house where Heidi and her grandfather stayed during the winter. We also visited traditional houses of farmers and climbed the mountain to observe the hut [see Figure 4.5] along the way. We saw many domestic Swiss trekkers there, and we talked and climbed the mountain with them. In addition, we took many photos of local people. (Bessatsu Takarajima, 2003: 21)

Kotabe also describes this part of the trip:

> we visited the hut and it was wonderful. Miyazaki used the physical appearance of the hut directly in the anime. In the location hunting, the translator negotiated and took us up to the hayloft. There, we saw straw and it looked so comfortable. We then used the image to create a depiction of Heidi's comfortable straw bed in the anime. (Bandai Visual, 2010: 11)

In addition to such sites, the tour group also visited art museums. Kotabe noted:

> It was the first foreign research trip for Takahata and me. In order to absorb the atmosphere of Heidi's world not only from the landscape but also from drawings, we visited the Segantini Museum in St Moritz to see the drawings by Giovanni Segantini, and the art museum in Chur to see those by Alois Carigiet, at Takahata's suggestion. (Kotabe, 2013: 86)

Kotabe also mentioned how before the location hunting he had designed the character of Heidi from his own imagination and with her hair tied in two pigtails. However, while in Switzerland, the director of the Johanna Spyri-Archiv said that her grandfather would not have tied the five-year-old Heidi's hair in pigtails. Heidi lived in a hut with only her grandfather and would not have spent much time on her hair. Kotabe was convinced by the director's arguments that Heidi would have had short hair, and this is how she ultimately appeared in the anime (Chiba, 2008: 81, 2017: 101–102; Kotabe, 2013: 86).

The group also gained information from materials found and purchased on site. Two books of Heidi with old illustrations that they bought in Zurich and picture books by Arois Carigiet, which were already published in Japan, helped with the descriptions of daily life (Takahata, 2008: 94). Watanabe also recorded the yodel voices of local singers during the location hunting and used them in the creation of the music for *Heidi* (Chiba, 2008: 89).

The location-hunting tour also contributed to the overall look of Heidi's world. Scene designer Hayao Miyazaki had visited Sweden in August 1971 as part of location hunting for an anime called *Pippi Långstrump*, although the anime was never made (Miyazaki, 1996: 565). This had been a good experience for him and informed his trip to Switzerland:

> I felt a good response through the location hunting in 1971 for the preparation of the anime *Pippi Långstrump*. [...] However, I had no particular image of Switzerland. [...] I tried to find out things within the scope of my abilities, for example, the architecture, mountains, vegetation, etc. [...] Actually, although we had many limitations, we did our best to create a respectable narrative world. (Bandai Visual, 2010: 5–6)

The group took many photos of local people and based the design of characters in the anime on them. This, director Isao Takahata believed, would make the characters more realistic (Bessatsu Takarajima, 2003: 21). There was also much sketching done. Kotabe said, 'I tried to sketch everything I saw', and mentioned that 'actually, the depiction of many kinds of characters were developed by these experiences' (Kotabe, 2013: 90). He also said that because it was his first time location hunting, he tried to take advantage of any and every opportunity to do something that would help in the creation of the anime as there were few materials in Japan that could be used (Bandai Visual, 2010: 9–10). The members of the group had different approaches to this process. Kotabe recalled, 'Miyazaki never sketched on site. Nakajima took photos, I sketched, but Miyazaki was just looking. However, after returning to Japan, Miyazaki started to draw what he had seen in Switzerland. I was really surprised to see it' (Chiba, 2008: 79).

However, all had used the trip to gain insights into Heidi's world which would never have been possible if they had not gone to Switzerland. As Kotabe noted, 'The space and light were completely different in Switzerland – more so than I had read in books and heard from others. While in Switzerland I felt, "Ah, this is where Heidi and the other characters live"' (Kotabe, 2013: 90).

Reconstructing the Heidi Narrative World and Cross-reference Tourism

As described in the previous section, the creators of *Heidi* successfully recreated the narrative world of Heidi by pursuing the expression of reality in fantasy, such as the 'layout system'. By doing so, they turned the narrative world of Heidi into one that many more people wanted to access as fans and tourists, thus imbuing Switzerland with a much more attractive tourism imaginary. Through this process of recreating and reediting a narrative world, or contentsization, as a matter of course the creators added new elements to the original contents. In contents tourism, cross-referencing the differences between the various media renditions of the contents (in this case, the differences between the original book and the anime) is very important to tourist attractions at related physical sites. Moreover, tourists also cross-reference differences between these media and the physical site itself (as media). As Beeton *et al.* (2013: 150) pointed out, the physical sites themselves 'become media sites in which meanings are transmitted from the creators'. In other words, the 'pre-existing' sites have been converted into media sites in which the narrative world of *Heidi* has been conveyed by both Spyri and anime creators to tourists. Therefore, tourists experience the narrative world in 'physical sites as media' and cross-reference differences between these sites and other media.

Through the cross-referencing process, the media (including 'physical sites as media') very often influence each other, and sometimes the real physical sites are changed when the contents are reconstructed by other media. *Heidi* is a typical example of this, and in this section the impact on tourism of the differences among these media renditions of the contents will be illustrated using three examples.

Another important aspect of this cross-referencing behaviour by tourists is travelling while checking the similarities and differences between *Heidi* and other artwork (in this case, drawings and picture books) that creators saw and were influenced by on their location-hunting tour. In the case of *Heidi*, the tourist experiences of the anime's creators have been mediatized and distributed in books, as interviews, and so on. Fan tourists can know whose artworks influenced the creators and want to cross-reference *Heidi* with those artworks. It can be said that the mediatized creators' experiences awakened the imaginations of fans into other works than *Heidi*. This aspect will be illustrated in the fourth example below ('Travelling to see works by Giovanni Segantini and Alois Carigiet').

Three fir trees

In the original Heidi novel, Spyri describes 'three old fir-trees with great shaggy branches' behind the hut:

> After three quarters of an hour they reached the height where the hut of the old man stood on a prominent rock, exposed to every wind, but bathed in the full sunlight. From there you could gaze far down into the valley. Behind the hut stood three old fir-trees with great shaggy branches. Further back the old grey rocks rose high and sheer. Above them you could see green and fertile pastures, till at last the stony boulders reached the bare, steep cliffs. (Spyri, 1919: Chapter I Going up to the Alm-Uncle)

In the original book, Spyri did not mention the location of the hut. However, since before the time of their location hunting, locals had called the hut the anime creators visited Heidi's hut and because it was located

Figure 4.5 Photo of a hut in Ochsenberg, Maienfeld, taken by Isao Takahata during the location hunting in July 1973. The hut became the model for 'the Alm hut' in *Heidi*.
Source: Photo courtesy of Isao Takahata.

in the area that the locals called 'Alm' (Chiba, 2008: 47; Chiba, 2017: 80). However, as indicated in Figure 4.5, there were no fir trees behind the hut when the creators visited it in 1973 (Chiba, 2008: 76–78).

However, the anime creators wanted to depict a symbol of Heidi's emotional support and decided to draw three fir trees in the anime just as described in the original book (Figure 4.6). Spyri placed Christian beliefs at the centre of the narrative world of Heidi in her original work and in the story Heidi's soul is saved via Christian belief. However, the anime creators gave the trees more important and symbolic meanings than in the original because, as Takahata commented, they decided to soften the Christian values in their anime version.

> I think Spyri wrote too much about Christian beliefs in her works. Although I sympathize with her goodwill, I carefully tried to soften the religious atmosphere and message based on Christianity, which is not very familiar to a Japanese audience. (Takahata, 2008: 96)

The other creators also perceived the same issue. According to Chiba (2017), Shigeto Takahashi (president of Zuiyo enterprise company, producer) and Isao Matsuki (screenwriter) thought that, in terms of Heidi's emotional support, specific religious belief should be replaced with something more symbolic. They considered the salvation of the soul in Christian belief to be not really accepted in non-Christian countries like

Figure 4.6 The image of the Alm hut in the anime *Heidi* (1974). © Studio 100 Media
Kindly supported by Studio 100 Media

Japan. They decided to soften the religious atmosphere and message based on Christianity, and instead of Christian beliefs they drew three fir trees behind the hut and gave them an important role as a kind of 'guardian encompassing the Alm' to them. They were also 'reflected in the anime screen' as the symbol of the Alm (Chiba, 2017: 70–71). Furthermore, the trees were modelled on a particular tree the creators saw on their trip to Switzerland. Kotabe said: 'near the Alm hut, we found a big, tall tree (during the location hunting) and I decided to use it as a model for the fir trees behind the hut' (Chiba, 2008: 78–79). Therefore, in the anime, Heidi 'listens to the leaves of fir trees rustle when she feels uneasy' and she 'recalls the fir trees at every opportunity when she is in Frankfurt' instead of praying to God (Chiba, 2017: 70–71).

The image of three large fir trees behind the hut is now broadly accepted as one of the most important and symbolic images relating to Heidi, not only among contents tourists but also local people. For example, the local tourism board uses the image for the cover page of its Heididorf brochure. Moreover, interestingly, locals planted three fir trees behind the hut which are growing there today (Figure 4.7). According to the Japan Heidi Juvenile Literature Meeting (n.d.), the trees were planted in 2004.

Peter and Heidi's adventure trail

The biggest difference between the anime and original version of Heidi was in the portrayal of Peter. Takahata felt that in the original

Figure 4.7 The same hut as in Figure 4.5 in Ochsenberg, Maienfeld, in August 2018. It is now called *'die original Heidihütte'* (the original Heidi hut) and used as a restaurant. Author's photo

version Peter was depicted compassionately as a humorous boy. 'This humour and compassion might have been based on the viewpoints of urban people looking down at a rural goatherd' (Takahata, 2008: 95–96). The creators wanted to depict him as 'an unsophisticated but healthy boy with virtues and good points, and as Heidi's good friend'. The creators felt that because the viewers had to keep company with him every week of the year on TV, they wanted the audience to feel that Peter was a friend. Consequently, they decided to change the original episode in which Peter jealously pushed Klara's wheelchair down the slope and it broke into a thousand pieces, despite that episode being one of the most memorable scenes in the original. 'So', Takahata commented, 'we made a new episode in which Klara accidentally broke the wheelchair by herself' (Bandai Visual, 2010: 2; Takahata, 2008: 95–96) (Figures 4.8–4.10).

Induced by the anime, many contents tourists now walk Heidi's Adventure Trail in Maienfeld, Switzerland, and find a signboard that says: 'the place where Klara's wheelchair broke into a thousand pieces after Peter pushed it down the slope' (Figures 4.11 and 4.12). They then cross-reference the anime and the original novel and many of them come to realize the differences between them. Such discoveries are a typical part of the contents tourism experience.

Figures 4.8–4.10 The scenes of Clara's accident in the 51st episode of the anime *Arupusu no Shōjo Haiji* (*Heidi, Girl of the Alps*), 1974. © Studio 100 Media Kindly supported by Studio 100 Media

76 Part 1: The Contentsization of Literary Worlds

Figure 4.11 A signboard for 'Peter's Place' on 'Heidi's adventure trail' (August 2018). Author's photo

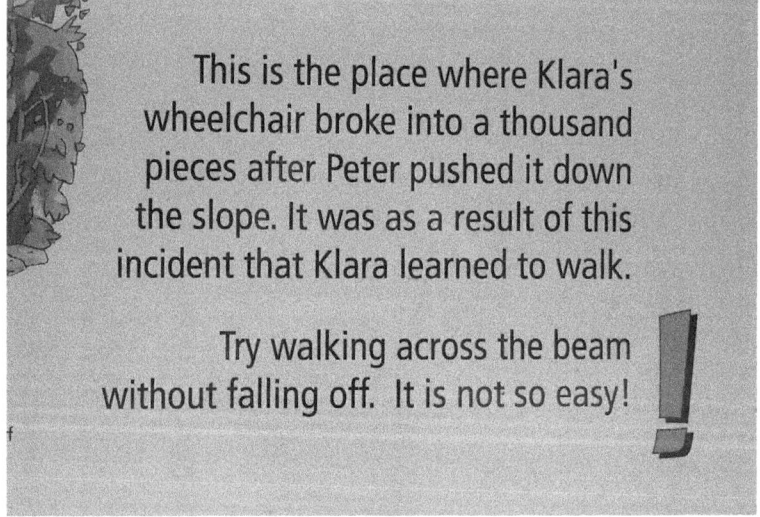

Figure 4.12 A part of the signboard for 'Peter's Place' (August 2018). Author's photo

Josef, the St Bernard

In the anime, there is a new character, Josef the St Bernard, who belongs to Heidi's grandfather and is not in the original book (Figure 4.13). Josef was created at the request of Shigeto Takahashi, the then-president of the anime production company and the producer (Bessatsu Takarajima, 2003: 21). After the anime, the image of Josef came to be widely accepted

Figure 4.13 Josef and Heidi in the fourth episode of the anime *Arupusu no Shōjo Haiji* (*Heidi, Girl of the Alps*), 1974. © Studio 100 Media Kindly supported by Studio 100 Media

by the local tourism industry, and was included on websites, brochures and souvenirs (Figure 4.14) together with the images of other animated characters that are also in the original. Takahata re-visited Switzerland in 2001 to make a presentation at an international conference on Spyri at which he commented that he was deeply impressed many Swiss people believe that the dog, the St Bernard, is Spyri's original character, although Josef was not in the original book (Bandai Visual, 2011: 6).

Travelling to see works by Giovanni Segantini and Alois Carigiet

As mentioned earlier, Takahata, Miyazaki and Kotabe visited Segantini Museum in St Moritz to see the drawings by Giovanni Segantini, and the art museum in Chur to see those by Alois Carigiet in order to 'absorb the atmosphere of Heidi's world' (Kotabe, 2013: 86). Kotabe's recollections of the period can be read in several articles. He said:

> 'Alpine Triptych: Life, Nature, Death,' the huge paintings by Segantini, were displayed in Segantini Museum, St Moritz. We saw them and felt they conveyed the atmosphere of Switzerland very strongly. For example, we felt as if we were experiencing the real cold of a Swiss winter when we saw the painting 'Death,' one of the 'Alpine Triptych.' We depicted images of winter in the anime by putting the image of the painting in place of a Japanese winter when we created anime *Heidi*. [...] In addition, I also used picture books by Alois Carigiet as a reference. When I in fact saw landscapes on site in Switzerland after reading these books, I could feel their atmosphere very vividly. (Bandai Visual, 2011: 8)

On Heidi Alp, with her gruff Alpöhi (her grandfather), Heidi spent the most beautiful time during the summers in her beloved mountains. To allow our visitors to enjoy this wonderful Alp experience up close, Heidi's Alp hut was reconstructed in great detail in Heididorf.

 *Heidi's Alp Hut with Alpöhi*

Heidi's Adventure Path

Heidi's Adventure Path from Heididorf to Heidi's Alp (Ochsenberg): 12 posts along the way tell the popular story of Heidi and Peter......

More information at.....

Heidi's Alp Adventure

.....www.heididorf.ch

Figure 4.14 The image of Josef on the official brochure of Heididorf, Maienfeld.
Source: Image courtesy of Heididorf

These descriptions show that the artworks of Segantini and Carigiet influenced *Heidi*. These mediatized creators' experiences awakened the imaginations of fans. By publicizing the experiences and influences that helped to create *Heidi*, fans can cross-reference the creators' experiences, these works of art and the anime itself. For example, Chiba compared Carigiet's picture books with *Heidi* and pointed out that some scenes of the books were absorbed into the anime. She mentioned that 'such a connection between Carigiet and *Heidi* can provide us with a richer experience of watching the anime *Heidi*' (Chiba, 2008: 65). This is a good example of fans' cross-referencing practices based on creators' experiences. In this way, St Moritz and Chur (and sometimes other sites associated with them, such as the town of Trun, the birthplace of Carigiet, where many of his wall paintings can be seen) can be also destinations for Heidi fans to see the artworks and to retrace the creators' experiences and feelings.

This is a representative example of the dynamic process of contentsization. That is, the contents (or 'narrative world') that connect specific sites/places in the minds of fans expand from the narrative world of the anime *Heidi* to that of the creators' experience, and furthermore to that of Segantini and Carigiet, through cross-referencing.

Conclusions

In this chapter, I have demonstrated how the creators' experience of transnational contents tourism helped them to reinterpret, recreate and remediatize the narrative world of Heidi in an example of the dynamic process of contentsization.

In past studies, not only on contents tourism but also on media-induced tourism, few articles have been written from a transmedia perspective, or about the creators' tourist experiences. However, the case study of *Heidi* clearly suggests that creators' experiences at locations as contents tourists have a strong impact on their recreation of a narrative world and act to strengthen the tourism imaginary. In other words, we can see the dynamic process by which the pre-existing sites were chosen as location models through the location-hunting tour and converted into media sites in which the recreated narrative world of Heidi and new elements were added to the original contents by anime creators, and transmitted to tourists. Therefore, in contents tourism, we can enjoy cross-referencing the differences between the various media renditions of the contents (in this case, the differences between the original book, the anime and the physical sites themselves) as important tourist attractions at related physical sites. Moreover, it was also observed that the mediatized creators' experiences awakened the imaginations of fans. Seeing the artworks that influenced the creators can be a reason for visiting Switzerland among contents tourists. It can be said that these cross-referencing behaviours are a particular aspect of tourist practices within the dynamic process of contentsization. Furthermore, based on the recreated narrative world, the cross-referencing process continues to be developed, and the *Heidi* case shows that there are even examples in which the real physical world has been reconstructed in accordance with the narrative world, as we saw in the examples of the three fir trees and Josef the St Bernard.

This chapter demonstrated, therefore, that analysing the dynamic process of contentsization while focusing on the cross-referencing process among the different mediatized formats and at tourist sites is an effective approach for clarifying the characteristics of contents tourism. This is a phenomenon that exists in many other important sets of contents disseminated across media formats and enjoyed in the real world by fans as tourists at related sites. Moreover, contentsization develops continuously through the cross-referencing practices of various contents tourists, which includes not only fans but also creators.

Acknowledgments

I would like to express my deepest gratitude to Mr Isao Takahata and Mr Junzo Nakajima for giving permission to use their photos, and to Mr Yoichi Kotabe, Mr Junzo Nakajima, Ms Kaori Chiba, Mr Tomosuke Suzuki, Mr Tetsuya Kawakami (Ghibli Museum Inc.), Ms Claudia Aebi (Heididorf, Maienfeld) and Ms Nicole Zaehringer (Johanna Spyri museum, Hirzel) for providing valuable information, making constructive comments on earlier drafts and helping to arrange photos and image permissions.

Notes

(1) The acceptance and popularity of Heidi by Japanese people has attracted the attention in German-speaking countries (German is the original language of Heidi). The German Wikipedia has a page about 'Heidi in Japan'. See https://de.wikipedia.org/wiki/Heidi_in_Japan (accessed January 2019).
(2) According to Mr Tomosuke Suzuki, Beta Film (Munich) broadcast the Japanese anime version of *Heidi* on TV in Spain, Italy, Austria and West Germany from the 1970s to the 1980s, and it became a big hit in each country. He said that they chose Spain first and Italy next, because black hair is more common in these countries within Europe. They were worried about whether European audiences could accept the black colour of Heidi's hair in the anime.
(3) The 'layout system' is one of the processes of creating Japanese anime works based on 'layouts', which are detailed design drawings. Creators conclude the compositions of backgrounds and the movements and postures of the characters by simulating a cut of the finished screen from storyboards drawn roughly, and then create a more detailed design drawing, or 'layout'. The system was first adopted officially in the creation process of *Heidi* (1974) by Hayao Miyazaki.
(4) The 'layout system' enabled Japanese animators to design and present screens with realistic architecture, such as distance perspectives, height, and depth. It is said to be a groundbreaking innovation that enabled Japanese anime to make their presentation of fictional stories highly realistic.
(5) I conducted interviews with Mr Nakajima at Keio Plaza Hotel Shinjuku, Tokyo, on 13 September 2017, and with Ms Chiba on 23 October 2017 at Cafe Trois Bagues, Jinbocho, Tokyo. I interviewed Mr Tomosuke Suzuki by telephone several times during September 2018.
(6) Information provided by Ms Claudia Aebi of Heididorf (personal communication, 27 September 2018). 'Heididorf' in German means 'Heidi village' in English. See http://www.heididorf.ch/en/english.html (accessed September 2018).

(7) It was the first foreign research trip for Takahata and Kotabe (Kotabe, 2013: 86), although others, such as Miyazaki, had been abroad before on a location-hunting trip. Before the creation of *Heidi*, Takahata, Miyazaki and Kotabe started working in 1971 on the plans of an anime version of the children's story *Pippi Långstrump* by Astrid Lindgren. In the same year, Miyazaki went to Sweden on a location-hunting tour for the plan (Kotabe, 2013: 86; Takahata, 1991: 163–165; Takahata *et al.*, 2014: 2, 64–72). Strictly speaking, this was the first location-hunting tour abroad in Japanese anime history. However, they did not obtain permission to create an anime version of *Pippi Långstrump* from Astrid Lindgren and the project was cancelled. The tour for *Heidi* in 1973 was the first location-hunting tour abroad that resulted in a TV anime series being created.

References

Bandai Visual (2010) Liner notes of *Arupusu no Shōjo Haiji* remastered DVD-BOX. Tokyo: Bandai Visual.
Bandai Visual (2011) Arupusu no Shōjo Haiji Interview Sairokushū. An appendix of *Heidi* Blu-ray Memorial Box. Tokyo: Bandai Visual.
Beeton, S., Yamamura, T. and Seaton, P. (2013) The mediatisation of culture: Japanese contents tourism and pop culture. In J.A. Lester and S. Caroline (eds) *Mediating the Tourist Experience: From Brochures to Virtual Encounters* (pp. 139–154). Farnham: Ashgate.
Bessatsu Takarajima (2003) Special interview vol. 1: Takahata Isao kantoku no 358 nichi. *Bessatsu Takarajima* 736, 18–21.
Chiba, K. (2008) *Arupusu no Shōjo Haiji no Sekai*. Tokyo: Kyūryūdō.
Chiba, K. (2017) *Haiji ga Umareta Hi: Terebi Anime wo Kizuita Hitobito*. Tokyo: Iwanami Shoten.
Hikawa, R. (2017) Nihon anime hyōgen shinkaron. *Geijutsu Shinchō* September 2017, 72–78.
Japan Heidi Juvenile Literature Meeting (n.d.) Haiji ten ripōto. See http://www.ne.jp/asahi/ts/hp/file5_heidi/heidi_text/5017_gibli_open.html (accessed May 2018).
Kotabe, Y. (2013) *Arupusu no Shōjo Haiji: Kotabe Yoichi irasuto gashū*. Tokyo: Kosaido.
Miyazaki, H. (1996) *Shuppatsuten: 1979–1996*. Tokyo: Tokumashoten.
Miyazaki, H. (2005) Panel for the exhibition *Arupusu no Shōjo Haiji ten: sono tsukuritetachi no shigoto*. Ghibli Museum, Mitaka, May 2005 to May 2006.
Okamoto, T. (2014) *Manga/anime de Ninkino 'Seichi' wo Meguru Jinja Junrei*. Tokyo: Xknowledge.
Okamoto, T. (2015) Otaku tourism and the anime pilgrimage phenomenon in Japan. *Japan Forum* 27 (1), 12–36.
Patton, M.Q. (2001) *Qualitative Research and Evaluation Methods* (3rd edn). Thousand Oaks, CA: Sage.
Spyri, J. (1919) *Heidi* (trans. E.P. Stork). Philadelphia, PA: J.B. Lippincott.
Takahata, I. (1991) *Eiga wo Tsukurinagara Kangaeta Koto*. Tokyo: Tokumashoten.
Takahata, I. (2008) TV series 'Arupusu no Shōjo Haiji' no haikei to sono seisaku wo megutte. In K. Chiba (ed.) *Arupusu no Shōjo Haiji no Sekai* (pp. 90–97). Tokyo: Kyūryūdō.
Takahata, I., Miyazaki, H. and Kotabe, Y. (2014) *Maboroshino 'Nagakutsushita no Pippi'*. Tokyo: Iwanami Shoten.
Wissmer, J.-M. (2015) *Haiji Shinwa: Sekai wo Seifuku shita Arupusu no Shōjo* (trans. T. Kawashima). Kyoto: Koyoshobo.
Yamamura, T. (2015) Contents tourism and local community response: Lucky star and collaborative anime-induced tourism in Washimiya. *Japan Forum* 27 (1), 59–81.

Part 2

Tourist Behaviours at 'Sacred Sites' of Contents Tourism

5 The Cotswolds and Children's Literature in Japanese Fantasy: The Case of Castle Combe

Catherine Butler

In the Introduction to this volume, Takayoshi Yamamura mentions the Japanese government's 2005 definition of contents tourism. Here, I would like to quote that definition at slightly greater length:

> We would like to call 'contents tourism' tourism that utilises content related to a local area (movies, TV dramas, novels, manga, games, etc.) to promote sightseeing and related industries.
>
> The basis of contents tourism is to add 'narrative' and 'thematic' qualities to a region – 'an atmosphere/image specific to the region generated by contents' – and to utilize that narrative quality as a tourism resource. (Ministry of Land, Infrastructure and Transport *et al.*, 2005: 49)

I quote this in full because, while the second sentence appears a natural extension of the first, potentially it carries quite a different emphasis. The first stresses the importance of specific movies, dramas and manga to contents tourism, but the second talks in more abstract language, of atmosphere (*fun'iki*) and narrative and thematic 'qualities' (*monogatarisei* and *tēmasei*); moreover, it speaks of *adding* these qualities to a place rather than exploiting what already exists there. In this chapter I want to explore the space between these two conceptions of contents tourism, suggesting that the ways in which Japanese tourists engage with the Cotswold region of England exemplify a range of interactions, some based firmly on specific content (for example, stories set in the Cotswolds), others more loosely on the region's power to catalyse the pre-existing desires and fantasies of tourists themselves through tropes and genre scripts.

Tourists arrive at any destination already equipped with what John Urry famously dubbed the 'tourist gaze': the 'particular filter of ideas, skills, desires and expectations' that frame and modify their experiences.

As Urry put it, 'when a small village in England is seen, what [tourists] gaze upon is the "real olde England"' (Urry & Larsen, 2011: 5). Similarly, readers have a repertoire of genre expectations that shape the interpretative approach they take to new texts. Indeed, these capacities may work in combination: literary scholars have noted 'the dominance and significance of certain aspects of the environment in the children's literature of different cultures' and the 'topographies regarded as typical in the representation of foreign landscapes' (O'Sullivan, 2005: 39) – associations that inform readings of individual texts. In this chapter, I will explore some of the ways in which place, contents and fantasy engage through Japanese tourism in the Cotswolds, and particularly in the small village of Castle Combe.

Japanese Tourism in the Cotswolds

The Cotswolds are a range of hills in southern England, lying mostly in Gloucestershire, Oxfordshire and Wiltshire. Historically important for sheep farming and the wool trade, the region grew rich in the Middle Ages before falling into relative decline in the 19th century, as the rise of cotton made wool less profitable. It was thus largely bypassed by the industrial revolution, and retains many traditional buildings built from its signature yellow (or 'honey-coloured') limestone. It remains a rural area, and is now popular with wealthy retirees and commuters seeking escape from urban life. Tourism is a major industry, and among the most enthusiastic tourists to the Cotswolds are those from Japan.[1]

Japanese tourists to Britain usually base themselves in London. When they leave the capital it is typically to visit places that can be easily reached from there in a day trip. The standard itinerary includes cities such as Oxford and Bath, and individual sites such as Stonehenge. But what of the countryside? Two rural areas are strongly promoted in Japanese tourist literature about Britain: the Lake District and the Cotswolds. Of these, the Lake District's status as a popular destination can be explained, in addition to the natural beauty of its landscapes, in 'classic' contents tourism terms. Its fame for Japanese people is bound up with Beatrix Potter and her most famous creation, Peter Rabbit; Potter's former home at Hill Top is a place of literary pilgrimage (Williams, 2013). The popularity of the Cotswolds at first seems more puzzling. The Cotswolds have no dominant literary locus like Potter's home at Hill Top or Shakespeare's at Stratford-upon-Avon. Writers, artists and other notable figures of history have of course lived in and written about the Cotswolds, but there is no individual for whose sake people visit in large numbers. No great events of history took place in the Cotswolds, no important battles were won or lost there. True, there are some small but historic cities at the edge of the area: Bath, Cheltenham, Gloucester, and slightly further afield, the larger cities of Bristol and Oxford – but when people think of the Cotswolds they do not generally picture these, but the picturesque villages and small towns and the countryside between them, none of them especially famous except for being pretty and for their bucolic names: Moreton-in-Marsh, Stow-on-the-Wold, Bourton-on-the-Water, Wotton-under-Edge, and so on.

Nevertheless, the Cotswolds have proved hugely attractive to Japanese tourists – to the extent that the railway station in Moreton-in-Marsh, where visitors from London arrive by train, has introduced bilingual signs for Japanese travellers (Figure 5.1). In some Cotswold villages – Bourton-on-the-Water, Bibury and Castle Combe among them – Japanese tourists at times constitute the majority of visitors, and this is reflected in the visibility of Japanese-language signage and Japanese-language souvenirs. There is no single reason for this popularity. One factor is no doubt the practical fact that the Cotswolds can be visited in a day trip from Paddington Station in London; several tourist guides in Japan include suggested itineraries for just such an outing (e.g. Diamond, 2017: 296). However, the Cotswolds are no more accessible from London than some other areas equally replete with beautiful countryside and picturesque villages. Some contingent circumstances may have played a part in cementing the Cotswolds' reputation in Japan. On 8 August 1890 William Morris (a figure long influential in Japan [Nakayama, 1996]) remarked in a letter to the designer Kate Faulkner that Bibury in the east Cotswolds was 'surely the most beautiful village in England, lying down in the winding

Figure 5.1 A sign in Japanese at Moreton-in-Marsh Station. Author's photo

valley beside the clear Colne' (Morris, 1996: 188). The claim of Bibury to be 'the most beautiful village in England' has been relentlessly leveraged ever since, not only by Bibury itself but by the region as a whole. This hyperbole has found its way into the representation of the region in material produced in Japan, where foreign tourist culture is characterized more than that of most Western countries by visiting approved destinations rather than on encouraging 'off-piste' exploration.[2] In 1921 the future Emperor Hirohito is said to have stayed in and praised Bibury as part of a six-month European tour, which may have further entrenched the region's position in the Japanese touristic image of Britain, even if most current-day Japanese visitors are unaware of the connection. More generally, the dominance of the Cotswolds is no doubt also a matter of success breeding success. The Cotswolds are prominent in Japanese tourist literature: numerous books are devoted to them specifically, and they punch well above their weight in general guides to Britain. In JTB Publishing's guide to the UK, for example, 20 pages are devoted to the Cotswolds, more than twice the number accorded to the whole of Scotland (JTB, 2017). Given this exposure, it is natural that the Cotswolds should form part of the itinerary of future visitors, too.

More relevant than these circumstantial elements to a discussion of contents tourism are the ways that the Cotswolds are used in imaginative and narrative terms – especially those associated with fairy-tale and children's literature. A recurrent theme of Cotswolds tourism for Japanese people is that in stepping into the Cotswolds one is stepping *out* of history, out of the modern world, into a place and lifestyle that have endured for centuries, and that in so doing one is entering a realm abstracted from the usual rules of time. Naturally, this is to a large extent an illusion; the area's appearance is the result of its being consciously preserved and maintained, not least for tourists. In some places, such as Castle Combe, byelaws ensure that such egregious signs of modernity as television aerials are not visible; in others, social pressure works to maintain a what is regarded as an appropriate appearance.[3] The Cotswold tourist experience is a result of a collaboration between the inhabitants and visitors in the creation of a selective vision of place, which excludes from view (literally, as far as brochures and postcards are concerned) anything that does not fit the narrative of rural timelessness. If the honey-coloured limestone and wisteria-fronted cottages of the Cotswolds are assiduously maintained by the residents and businesses who depend on tourists for their prosperity, they are no less curated by the tourists who come to see them.

The question remains, what use is made of this perceived quality of timelessness? How is it processed and understood by visitors? For some visitors, it seems that stepping out of history offers an escape from everyday worries. David Strachan, proprietor of Totteoki Cotswolds Tours, a small company that offers Japanese visitors personal tours of the Cotswolds, recalls:

> One woman ... there was a couple and their teenage daughter. We met them at the station. They'd come from Japan to London, London to Moreton, then into our vehicle. We took them to a little village. She got out and burst into tears, because, she said, she didn't realise anywhere could be so beautiful. She was like that the rest of the day, overwhelmed by everything.
>
> Another mother and daughter, at the end of the tour I asked them what they thought of it, and the mother said, 'I feel like I've been cured.' And then she went all wet-eyed, and her daughter said she'd been under a lot of stress, she lived in Tokyo, and her job and family situation was really stressful, and spending the day just chilling out in the Cotswolds just made it all go away. (Strachan, 2017)

The reality of Tokyo 'goes away'; but what takes its place? I suggest that the experience of the Cotswolds is to a large extent filtered through Japanese visitors' knowledge of, and associations with, children's literature. My own encounters with Japanese tourists in the Cotswolds have typically elicited comparisons with picturebooks (*ehon*) and nursery stories (*otogibanashi*), as well as more specific children's texts such as *Alice in Wonderland*, *Peter Rabbit* and Harry Potter. Being in the Cotswolds allows visitors to interpolate themselves imaginatively into this kind of fictional environment. This is not contents tourism as conventionally understood: none of these texts is set in the Cotswolds.[4] Nevertheless, with their traditional architecture, rurality and lack of obvious markers of modernity, Cotswold towns and villages evoke a certain aesthetic strongly associated with fairy tales and strikingly distant from the daily experience of most visitors from Japan, where few domestic buildings date back as much as a century. Moreover, an image of rural England is absorbed by Japanese people at a young age, not only through literature but also through other aspects of children's culture. The Sylvanian Families toys, for example, developed by the Japanese company Epoch and first sold in 1985, were designed to evoke a rural English, middle-class lifestyle much like that of the Cotswolds. With its animal characters, they also recall a childlike, diminutive world of the kind Beatrix Potter created in her stories.

Size is an important part of this aesthetic. Stroud and Cirencester, the largest towns in the Cotswolds, have populations of 32,000 and 19,000. The other settlements are all much smaller. Bourton-on-the-Water has a population of just 3300, Bibury less than 700. Individual buildings are often built with low doors and ceilings designed for smaller-framed generations. This element of 'snugness' is another element associated with children's texts. As Jerry Griswold has noted, 'Only in children's literature is littleness so frequent a topic ... and only in that genre does the word *little* appear so frequently in titles' (Griswold, 2006: 51). The Cotswolds offer a smallness that allows visitors to indulge aspects of the imagination that evoke fairy tales and children's stories about beings, small and large.

When the Japanese produce Cotswold-inspired attractions in Japan, such as Yufuin Shopping Village in Oita Prefecture, or Dreamton near Kameoka in Kyoto Prefecture (of which more below), there is typically an emphasis on the smallness and 'cosiness' of the environment, which offers a bespoke pleasure to child visitors and a regressive one to adults.

Three Types of Contents Tourism in Castle Combe
Kin-iro Mosaic and 'traditional' contents tourism

The appeal of the Cotswolds, in the general terms in which I have described it here, fits rather imperfectly the first part of the definition of contents tourism quoted at the start of this chapter, as 'contents related to the local area (movies, television dramas, novels, manga, games and so on)'. The Cotswolds has no Beatrix Potter, nor even Harry Potter. Yet the region is flexible and effective in terms of the second part of that definition, the addition of narrative and thematic qualities, and of an atmosphere that can be exploited as a resource for contents tourism. The Cotswolds' power to evoke narrative possibilities in visitors who come equipped with the imagery of literature and culture of childhood, makes it a potent contents tourism location.

Some of the ways in which children's-literature-related contents tourism manifests in the Cotswolds can be seen by turning from a region-wide perspective to a more specific location. Castle Combe in the south Cotswolds is a small (population 350), famously pretty village, in which almost all the houses are built from Cotswold stone and are several centuries old (Figure 5.2). In the words of one Japanese guide book, its 'rows of stone houses seem to have been taken from a fairy-tale' (Kobayashi & Saito, 2015: 124). Castle Combe is on the regular itinerary of many Japanese tourists, particularly those coming from nearby Bath. Public transport links being sparse, Japanese visitors generally come as part of a tour, which may include a visit to a village pub, tea room or the local manor house. Linguistic interactions are basic, with few or no residents being fluent in Japanese, and most Japanese having minimal English.

Castle Combe has been used as a television and film set on many occasions. Among other roles, it was transformed into the harbour of Puddleby-on-the-Marsh in the 1967 musical version of Hugh Lofting's *Dr Dolittle* books, and the English village in Steven Spielberg's 2011 film of Michael Morpurgo's First World War novel, *War Horse* (1982). One might expect that tourism would be generated by such appearances in television and film, but, despite brief surges of interest at the time of the films' release, this kind of film-induced tourism accounts for only a small proportion of visitors. In this respect Castle Combe can be contrasted with nearby Lacock, a larger village just outside the Cotswolds, which openly trades on its many appearances in television and film, ranging from

The Cotswolds and Children's Literature in Japanese Fantasy 91

Figure 5.2 Castle Combe, Packhorse Bridge. Author's photo

Pride and Prejudice to *Downton Abbey* and *Harry Potter*.[5] Nevertheless, the Cotswolds in general and Castle Combe in particular have benefitted from the Japanese custom of anime tourism. Given that the Cotswolds are an area of England well known in Japan, it is not surprising that several manga and anime have been set there, although the word 'Cotswolds' is not always used.[6] The most relevant example to Castle Combe is *Kin-iro Mosaic*, a television anime aired by Studio Gokumi from 2013 and based on the four-panel manga by Yui Hara (2010). The story is largely set in a Japanese high school, but in the first episode ('In Wonderland') Shinobu, a middle-school student, visits England for a homestay and becomes friends with Alice, an English girl of the same age. After Shinobu returns home, Alice learns Japanese in England and transfers to Shinobu's school to join her and her friends. *Kin-iro Mosaic* tells of their life together in Japan, but frequently refers back to Alice's home in the Cotswold countryside.

In the manga, Alice's home is depicted as a wooden-walled building set amid snow-topped mountains more redolent of Japan than the Cotswolds (Hara had not at that point visited the UK), but Studio Gokumi sent a team to photograph specific locations, modelling Alice's house on Fosse Farmhouse, a guesthouse just outside Castle Combe (see also Yamamura, Chapter 4); other Cotswold locations included Cirencester, Bibury, Kemble Station and Bathampton. Even the address of Fosse Farmhouse ('Fosse Farmhouse, Castle Combe, England') can be seen on a pot of homemade jam briefly glimpsed on Shinobu's Japanese breakfast table in the second episode ('Even if I'm Small'). These references naturally generated interest among fans, and Caron Cooper, the proprietor of Fosse Farmhouse, found that the number of her Japanese visitors increased dramatically after the anime was broadcast in 2013. By 2017 she estimated that some 70% of her overnight guests were Japanese, almost all attracted by the anime, and many including *Kin-iro Mosaic* as part of an anime pilgrimage around Britain or even Europe (Cooper, 2017). One room in Fosse Farmhouse is now a *Kin-iro Mosaic* shrine, featuring merchandise and signed photographs and messages from famous visitors, including the voice actresses and Yui Hara herself (Figure 5.3). Visitors practise most of the forms of *otaku* behaviour Takeshi Okamoto has described as characteristic of anime tourism (Okamoto, 2015: 24–26): taking pictures of locations from the same angles as the anime, re-enacting scenes, creating photograph albums in which a figurine or doll substitutes for the anime character, and so on. Because visitors generally want to see the house exactly as it was shown in the anime, Cooper finds it difficult to change any of the ornaments, the bed spreads, and so on. In this sense, Fosse Farmhouse has become as fixed in 2012 (when the studio photographers came) as the characters in the anime itself.

Anime contents tourism now generates a significant proportion of Fosse Farmhouse's revenue, but there is none of the co-ordination with public tourism bodies that has become common in Japan, where anime studios and city and prefectural governments often work together to

Figure 5.3 *Kin-iro Mosaic* shrine at Fosse Farmhouse. Author's photo

co-promote an anime and its setting, resulting in promotions such as the Anime Tourism Association's 88-Stop Pilgrimage (Anime Tourism 88, 2018). Cotswold tourism officials have done nothing to foster this market, seeing it as too niche to warrant the use of their limited resources (Jackson, 2018). Until recently, indeed, they were unaware of this kind of tourist activity, although that situation may be changing; in April 2018, Fosse Farmhouse won an award for the 'Story of the Year' from the Automobile Association on the basis of its anime connection (Automobile Association, 2018), and in early June the story was taken up by most British newspapers and news websites, even briefly topping the 'most viewed' item ranking on the BBC's news website (BBC, 2018). Clearly there is a potential appetite in Britain for anime tourism, at least as a human-interest story.

Do-it-yourself contents tourism

Kin-iro Mosaic notwithstanding, contents tourists constitute only a small percentage of Japanese visitors to Castle Combe. Most of those who come by coach or minibus as part of a tour do so not for the sake of a specific narrative but for the village's rustic beauty. Nevertheless, at least one business owner has attempted to grow her sales by creating a kind of do-it-yourself contents tourism experience. Anna Roberts sells cream teas from the Old Rectory Tea Room in Castle Combe (her shop is featured in some Japanese tourist guides; e.g. Kobayashi & Saito, 2015: 125), but also uses her premises to sell souvenirs. These include knitted toy mice, which are popular with Japanese visitors. A toy is just a toy, but a toy with a story is more powerful, and a toy with a story set in a real place more powerful still. Accordingly, in 2017 Roberts self-published a book about the mice and Castle Combe, entitled *Mouse Tails of Castle Combe.*

Mouse Tails tells of a family of mice, the Whiskerfords, who are tired of living in London and long to move to the country. Eventually they arrive in Castle Combe and take up residence in the local tea room. This is a fantasy for tourists, but it is also the story of the author's own family, as she explains on the back cover:

> In 2003, Anna and Mike Roberts moved into the Old Rectory house here in Castle Combe, where they now run a Tearoom from the ground floor of their private family home.
>
> This is a story about them, told through a family of mice called 'The Whiskerfords.'
>
> All the characters in the book can be seen displayed in the tearoom windows and are based on real life villagers. (Roberts, 2017: Back Cover)

By telling her own story through a family of mice, Roberts takes quite a typical story (a London couple buy a house in the country) and filters it through a long tradition of children's literature, of animal stories, and even more specifically of mouse stories (the ancient story of the Town Mouse and the Country Mouse is one obvious precursor). She thus infuses the place with 'narrative quality', both for herself and potentially for her readers. Roberts has even constructed a mouse front door on her staircase, to pique the interest of visitors to her tea room. Copies of the door are, of course, for sale in the shop, along with the mice and the book itself. After all, this is not just a fantasy, it is also a business, and the creation of these contents allows different aspects of that business to be integrated through narrative. For example:

> They carried on all the way down the street until they reached the Rigglesby's Tearoom. The heavenly smell of freshly baked scones came drifting out of the half-open stable door and poured out onto the street. (Roberts, 2017: n.p.)

In this passage and others like it, the village, the book, the toys and even the cream teas all combine to promote each other.

Mouse Tails is as much a tourist guide as it is a story, and it contains many soft-focus pictures of the various buildings in the village. Although Japanese visitors may not be able to read the English text fluently, the book potentially functions as an *omiyage* (or souvenir) to remind them of Castle Combe. Anime studios may not yet be cooperating with local government in the Cotswolds, but here a similar strategy is being carried out at an individual level: this is contents tourism grown from seed.

Stay-at-home contents tourism

Not every Japanese person with an interest in Britain has the resources or even the desire to travel there. To cater to such people, a number of places have been constructed in Japan that recreate, or at least gesture towards, a certain vision of Britishness. These include British Hills in Fukushima, a village complex used primarily for educational purposes, and tourist attractions such as Yufuin Floral Village in Oita Prefecture. Most relevant to Castle Combe is Dreamton near Kameoka, in Kyoto Prefecture (Figure 5.4). Dreamton is the creation of Mayumi ('Marie') Haruyama, the Anglophile scion of a family of traditional weavers from Nishijin, Kyoto. Inspired by the hospitality and way of life she encountered on a visit to the Cotswolds, Haruyama created Dreamton in the early 2000s, in tribute to the aesthetic, craft and social values of the region. The name 'Dreamton' is of course significant, marking it as a place that can be used to give dreams solid form – whether those of the owner or of visitors. Dreamton comprises a restaurant, a row of bed-and-breakfast cottages, a chapel for weddings, an antique shop and various other buildings in the Cotswold style. The distinctive honey-coloured Cotswold stone had to be reproduced using concrete, but the effect is convincing, and the interiors are largely furnished with British items purchased by

Figure 5.4 Dreamton. Author's photo

Haruyama during her regular antique-dealing trips to the UK. Notably, the image used to advertise the Pont-Oak restaurant, which was the first stage of the project, is a picture of Castle Combe, a place that has in some ways served as an archetypal model of a Cotswold village.

In interview, Haruyama stressed that she does not wish Dreamton to be a theme park (Haruyama, 2018), but rather as far as possible a recreation, allowing visitors an immersive experience in which they feel themselves to be truly in England. In the case of Dreamton, the village of Castle Combe isn't just associated with existing contents (an anime, a children's book), it *is* the contents – a place designed to reproduce the same associations, the same atmosphere and thematic qualities as the Cotswolds themselves.

Conclusion

In the Cotswolds, the absence of strong associations with specific works or authors, of the kind exemplified by Beatrix Potter in the Lake District, limits the possibilities of one familiar kind of contents tourism, but it opens up the space for a more creative use of the region's landscape. The Cotswolds are a flexible imaginative space, in which both tourists and those who curate tourist sites make use of generalized imagery and narrative associations of folktale, nursery stories and traditional children's literature, and project them onto the Cotswolds, idealizing that region as a site for such contents: rural, set slightly back in time and slightly to one side of history. Within this environment, I have identified three examples of contents tourism associated with the small village of Castle Combe. Each is associated not with a major industry but with a small enterprise owned and promoted by an individual businesswoman: Caron Cooper, Anna Roberts and Marie Haruyama. It is easy for such small-scale enterprises to fly beneath the radar of large tourism organizations and of academic notice, but in fact this literal 'cottage industry' is typical of the scale at which tourism in the Cotswolds operates. Its ingredients include the material reality of the Cotswolds themselves, the fantasies and expectations of the people who visit, and the ways in which these are shaped by children's literature and other existing cultural structures. The varying types and degrees of relationship these enterprises have to narrative contents suggests that the addition of 'atmosphere' and 'thematic qualities' may in many cases be far more than a way of exploiting or adding to existing contents; it may also constitute those contents.

Acknowledgment

The field research for this chapter was funded in part by a research grant from the Japanese Foundation Endowment Committee.

Notes

(1) There are no official figures for the number of Japanese visitors to the Cotswolds, but an informal survey of retail business owners and staff in Tourist Information Offices across the region suggests that they make up a significant proportion of tourists, and that in those towns and villages particularly promoted in Japan (notably Bourton-on-the-Water, Bibury and Castle Combe) may be the most numerous foreign visitors.
(2) Morris's words echo even as far as the website of the Hotel Monterey Grasmere in Osaka, the 22nd floor of which boasts a reproduction of Brockhampton Church in Herefordshire, which is used for wedding ceremonies. According to the hotel website, 'The design imitates the churches of the Cotswolds, described by the renowned designer William Morris as the most beautiful in England. Our chapel recreated the beauty of the Cotswolds, from its rolling green hills to its traditional arts and crafts culture'. This, despite the fact that Morris was not discussing churches, and that Brockhampton (though designed by an Arts and Crafts architect) lies some 30 km outside the Cotswolds (Hotel Monterey Grasmere Osaka, n.d.).
(3) The case of a yellow car in Bibury, vandalized because it was thought to spoil tourists' photographs, is one noteworthy example of recent times (BBC, 2017).
(4) J.K. Rowling was born and brought up near the Cotswolds, but never lived there herself nor set any scenes from the Harry Potter books there. That said, parts of *Harry Potter and the Chamber of Secrets* (2002) and *Harry Potter and the Half-Blood Prince* (2009) were filmed in nearby Lacock, which is built in the Cotswold style and is regularly included in the Cotswold itineraries followed by Japanese tourists.
(5) For example, Lacock boasts a Harry Potter-themed shop, and books such as *The British Television Location Guide* (Clark, 2013) are on prominent display at the village's National Trust shop.

(6) Kore Yamazaki's *The Ancient Magus' Bride* (Yamazaki, 2013) is an example of a story in which both manga and anime make use of Cotswolds settings (including Broadway, Burford and Bourton-on-the-Water), without ever naming any of these locations.

References

Anime Tourism 88 (2018) Anime pilgrimage sites. See https://animetourism88.com/en/sanctuary (accessed July 2018).
Automobile Association (2018) Britain's best B&Bs announced at the AA B&B Awards. See https://www.theaa.com/hospitality-awards/news/bed-and-breakfast-awards-2018 (accessed October 2018).
BBC (2017) Notorious yellow car vandalised in Bibury. 4 February. See https://www.bbc.co.uk/news/uk-england-gloucestershire-38867290 (accessed July 2018).
BBC (2018) The English B&B that is famous in Japan, 5 June. See https://www.bbc.co.uk/news/uk-england-wiltshire-44368853 (accessed October 2018).
Clark, S. (2013) *The British Television Location Guide*. Bishop's Waltham: Splendid Books.
Cooper, C. (2017) Interview with the author, 12 April.
Diamond (2017) *Chikyū no Arukikata*. Tokyo: Diamond.
Griswold, J. (2006) *Feeling Like a Kid: Childhood and Children's Literature*. Baltimore, MD: Johns Hopkins University Press.
Hara, Y. (2010) *Kiniro Mozaiku [Kin-iro Mosaic]: Manga Time Kirara Max*. Tokyo: Hōbunsha.
Haruyama M. (2018) Interview with the author, 19 May.
Hotel Monterey Grasmere Osaka (n.d.) Facilities. See https://www.hotelmonterey.co.jp/en/grasmere_osaka/facilities/ (accessed July 2018).
Jackson, C. (Shared Tourism Officer at Cotswold and West Oxfordshire District Councils) (2018) Interview with the author, 13 April.
JTB (2017) *Igirisu: Tabimoto Ōshū 10*. Tokyo: JTB.
Kobayashi, S. and Saito, R. (2015) *Kottsuworuzu: Ingurando de Ichiban Utsukushii Basho*. Tokyo: Diamond.
Ministry of Land, Infrastructure and Transport, Ministry of Economy, Trade and Industry and Agency for Cultural Affairs (2005) Eizō tō kontentsu no seisaku, katsuyō ni yoru chiiki shinkō no arikata ni kansuru chōsa. See http://www.mlit.go.jp/kokudokeikaku/souhatu/h16seika/12eizou/12_3.pdf (accessed July 2018).
Morris, M. (1996) Letter 1743. In N. Kelvin (ed.) *The Collected Letters of William Morris, Volume III, 1889–1892* (p. 188). Princeton, NJ: Princeton University Press.
Nakayama, S. (1996) The impact of William Morris in Japan, 1904 to the present. *Journal of Design History* 9 (4), 273–283.
Okamoto, T. (2015) Otaku tourism and the anime pilgrimage phenomenon in Japan. *Japan Forum* 27 (1), 12–36.
O'Sullivan, E. (2005) *Comparative Children's Literature* (trans. A. Bell). London: Routledge.
Roberts, A. (2017) *Mouse Tails of Castle Combe: The Whiskerfords*. Castle Combe [self-published].
Strachan, D. (2017) Interview with the author, 9 October.
Urry, J. and Larsen, J. (2011) *The Tourist Gaze 3.0*. London: Sage.
Williams, F. (2013) Peter Rabbit: Why the Japanese love Beatrix Potter. See https://www.bbc.co.uk/news/uk-england-cumbria-24625202 (accessed July 2018).
Yamazaki, K. (2013) *Mahōtsukai no Yome [The Ancient Magus' Bride]*. Tokyo: McGarden.

6 Yōkai Tourism in Japan and Taiwan

Shinobu Myoki

Yōkai are mysterious creatures in Japanese folklore. Originally, they were invisible, but people have given them appearances, names and interpretations. Yōkai images, therefore, have been created and consumed as works of popular culture. There has been a boom in exhibitions dealing with yōkai, the earliest of which was a picture exhibition held at the Hyōgo Prefectural Museum of History in 1987 (Komatsu, 2003: 26–27). Originally, yōkai were intended to terrify, although sometimes they are depicted as cute, as indicated by the 'From Eerie to Endearing: Yōkai in the Arts of Japan' exhibition held in Tokyo and Osaka in 2016. Komatsu (2006: 10–17) states that the concept of yōkai can be divided into three semantic domains: yōkai as event or phenomenon, yōkai as mysterious presence and yōkai as visualized object (*zōkei*). Michael Dylan Foster (2015: 8), meanwhile, notes that 'Yōkai dwell in the contact zone between fact and fiction, between belief and doubt'. Yōkai are a presence produced by the human mind, and therefore, to study yōkai is to study the human mind and human society (Komatsu, 2007: 154, 279).

In this chapter, I will discuss Kai ('mystery') heritage and yōkai at three sites in Japan, as well as examining the use of yōkai at a theme park in Taiwan.[1] In the first section, I will clarify the characteristics of each site in order to identify similarities and differences with respect to yōkai tourism. I will then analyse how yōkai function in their community, focusing especially on Yamashiro. This second section is subdivided into two parts: first, how residents interpret yōkai in their community and daily life, and how they protect yōkai legends; second, how activities among residents can lead to yōkai tourism and how yōkai tourism functions (between yōkai and tourists and between residents and tourists) through culinary activities and children's activities. However, during my fieldwork I learned that, through the invisible presence of yōkai, the activities of Yamashiro's residents brought about international exchange. In the third section, therefore, I discuss the communication between Xitou Monster Village in Taiwan and Yōkai Village in Yamashiro. In the conclusion, I discuss how yōkai enable movement across time and place, and their status as a tool for enabling creative activity. I also examine the way yōkai images/icons are consumed within and beyond their original narratives. In so doing, the role that yōkai tourism plays in contemporary Japan and Taiwan is revealed.

Kai Heritage in Japan: Sakaiminato, Tōno, Yamashiro

As of 2018, there are three Kai heritage areas in Japan. The *Sekai Yōkai Kyōkai* (World Yōkai Association, whose first and honorary posthumous president is Shigeru Mizuki) designated Sakaiminato City in Tottori Prefecture (western Honshū) in 2007, Yamashiro Town in Tokushima Prefecture (Shikoku) in 2008, and Tōno City in Iwate Prefecture (northern Honshū) in 2010 as sites of Kai heritage.[2] Sakaiminato is the hometown of Shigeru Mizuki (1922–2015), who created the yōkai manga *GeGeGe no Kitarō*. Tōno is the location of *Tōno monogatari* (The Legends of Tōno) written by Kunio Yanagita in 1910. Yamashiro, meanwhile, is the birthplace of the Konaki-jiji (a yōkai with the face of an old man and the voice of a crying baby) legend. These three sites share similarities as destinations for yōkai tourism, but there are also differences in their historical backgrounds and the contents that tourists attempt to find through tourism to each area.

Sakaiminato is famous for Mizuki Shigeru Road, which is lined with bronze statues of yōkai. According to Sakaiminato City Tourism Association (2018), the opening ceremony was held on 18 July 1993, at which time there were 23 statues in total. Since then, the number of statues has gradually increased, reaching 177 by the time the road reopened after renovations on 14 July 2018 (the most recent statues being added on 28 June that year). The Mizuki Shigeru Museum also opened in 2003 (see Figure 6.1). In 2010, Mizuki Shigeru Road had 3,724,196 visitors and Mizuki Shigeru Museum had 411,006 (Sakaiminato City, 2017). This was a larger number than usual owing to the broadcast of *GeGeGe no Nyōbō* (an NHK morning soap opera about Shigeru Mizuki's wife, Nunoe Mura) on television that year.

In Takayoshi Yamamura's categorization of the patterns of anime/manga tourism, one pattern is places with a connection to the original author of the work. Sakaiminato can be considered one of these (Yamamura, 2011: 24–25). At Shōfukuji Temple, which Shigeru Mizuki visited in his youth, there is a famous picture of heaven and hell, and these pictures are considered to be the origin of his yōkai sketches. Sakaiminato possesses not only these historical contents and many stories and statues of yōkai from all over Japan, but also elements related to manga, anime and television. Tourism in Sakaiminato, therefore, is a good example of contents tourism. According to Yukio Shōji (2018), Director of the Mizuki Shigeru Museum, the reason that Mizuki's contribution is so great is that he drew many yōkai which were originally known only to local people. He popularized yōkai to the public in Japan and around the world (Shōji,

Figure 6.1 International tourists and someone in a Sunakake-babaa yōkai costume in front of the Mizuki Shigeru Museum (13 October 2017). © MIZUKI Production. Author's photo

2018). Yōkai now form a key part of local identity. For example, in 2013 the Mizuki Shigeru Museum published educational materials for children titled 'Let's get along with the yōkai in *GeGeGe no Kitarō*!' (Mizuki Production & Imamura, 2013). This heritage project teaches children about yōkai culture and their hometown's attractions.³

The second site of Kai heritage, Tōno, is the setting of the Legends of Tōno. Tōno has a famous oral tradition in which *kataribe* (storytellers) begin their stories with the words '*Mugasu attazumona*' ('They say, a long time ago …') and end them with '*Dondohare*' (there are several interpretations of its exact meaning, but it is used to conclude a story).⁴ These narrative conventions and the stories themselves are passed down to younger generations. In 2009, the Tōno *Kataribe* 1000 People Project was organized by Tōno City. Storytellers are not only adults, but children, too. The Tōno City government collaborates with elementary schools in the city and children learn about tales of the past (*mukashibanashi*). For example, on most weekends and holidays from May to November, students from Tsuchibuchi elementary school, accompanied by an adult storyteller (see Figure 6.2), take turns to tell tales of the past at Denshōen, a park and exhibition complex built to protect Tōno's traditional culture and way of life and pass it on to future generations (Denshōen, 2010).

Tōno's Kappa-buchi is a famous pool, which many tourists visit to find kappa (a kind of yōkai that inhabits aquatic areas such as rivers). Haruo Unman, whose nickname is Kappa Ojisan (Uncle Kappa), can sometimes

Figure 6.2 A Tsuchibuchi elementary school student tells the story of 'Kappa no Ongaeshi' (the Kappa repays a favour) at Denshōen (22 July 2018). Author's photo

be found there providing commentary to tourists. Unman is considered a *maburitto*. In the Tōno dialect, *maburu* is a verb meaning 'protect' and *maburitto* is the noun derived from the verb. In this context, a *maburitto* is a person who conveys and protects Tōno's folklore. Unman also explains the history and folklore of Tōno to tourists at the Denshōen. Unman says that he performs his role as a *maburitto* in order not to let the wisdom of his predecessors go to waste (Unman, 2018; see also Figure 6.3). In this way, not only the stories and the landscape, but also the inhabitants of Tōno become important elements of Tōno's contents tourism.

Nowadays, another form of the visualization of yōkai is stamp rallies, in which people visit suggested places and collect stamps from each one. For example, a stamp rally from 2016 reads, 'Get prizes for each yōkai stamp that you collect. Find the yōkai in Tōno. Can you find all of them?' The yōkai in this stamp rally were depicted as cute (Figure 6.4). This shows a different tendency from that in Sakaiminato and Yamashiro. The former's bronze statues and the latter's yōkai dolls are frightening rather than cute.

The third and final site is Yamashiro, located in central Shikoku, the smallest of Japan's four main islands. It is famous for its beautiful natural surroundings, including the Ōboke and Koboke gorges. Yamashiro Town is covered with steep mountains, and this precipitous landscape is of direct relevance to the widespread persistence of yōkai folklore (Hirata, 2018a). That is, these stories are needed to protect people, especially children, from natural dangers. In 1938–1939, Kunio Yanagita published the

Figure 6.3 Maburitto Haruo Unman explains the history and folklore of Tōno to tourists in Denshōen (22 July 2018). Author's photo

Yōkai-meii, a glossary that includes 80 kinds of famous Japanese yōkai. Shōichi Shimooka (Shimooka, 2009: preface) notes that the *Yōkai-meii* includes four types of yōkai spoken of in present-day Yamashiro. They are Konaki-jiji (Yanagita 1938a: 12), Takabōzu (Takanyūdō in Yamashirodani) (Yanagita, 1938b: 12), Yagyōsan and Kubinashiuma (Yanagita, 1939: 16).[5] Yanagita (1938a: 12) wrote that Konaki-jiji is a yōkai found in villages in the mountainous regions of Awa province (present-day Tokushima Prefecture).

A local historian from Anan City, Masahiro Takita, began his research on Konaki-jiji's origins in 1998 (Tokushima Shinbun, 1999). He conducted surveys in about 20 cities, towns and villages in the mountainous regions of the prefecture, and during his research, he found a reference written by Akira Takeda (Tokushima Shinbun, 1999). Takeda (1938: 10) had identified the name Konaki-jiji, provided an explanation of it and attributed it to 'Sanmyō-son [village] Aza Taira' (present-day Taira in Kamimyō, Yamashiro Town). Takita found three people who had heard the Konaki-jiji legend (Hirata, 2018a; Takita, 2002a).[6] In 1999, he confirmed that its origin was the Kamimyō area of Yamashiro Town (Takita, 2002b).

In this way, Konaki-jiji, which originally came from present-day Tokushima Prefecture, was later visualized in manga by Shigeru Mizuki, then adapted into popular culture, such as anime series. It was after the story of the Konaki-jiji was popularized in this way that its origins in Yamashiro Town were rediscovered by Takita. Hiroya Ichikawa (2013a: 130–133) discusses the process through which yōkai perspectives are

Figure 6.4 Yōkai stamp rally card, 2016 (reverse side) (one stamp affixed by the author). Received from Tōno Tourist Association. Event organized by the Tōno City SL Teishaba Project Promotion Committee. Used with permission of the Committee office at Tōno City Hall

formed: legends, conversion to text, visualization, the propagation of images thorough manga, the excavation of yōkai folklore and becoming resources for regional regeneration. Ichikawa (2013a: 136) points out that in Yamashiro there are at least two yōkai perspectives: one is yōkai as characterized in popular culture, and the other is yōkai as folkloristic culture in individual memories.

After Takita's contribution, in 2001, a stone statue of Konaki-jiji was built in Yamashiro. In 2008, Shikoku-no-hikyō Yamashiro Ōboke Yōkai Mura (Yamashiro Ōboke Yōkai Village in Shikoku's Secluded Areas, hereafter 'Yamashiro Ōboke Yōkai Village') was established.[7] In the same year, Yamashiro Ōboke Yōkai Village was certified as a Kai heritage site. In December 2008, Yamashiro Ōboke Yōkai Village unveiled a development plan (Yamashiro Ōboke Yōkai Mura, 2008: 3), declaring that they would make use of regional resources such as history, legends (yōkai), residents (storytellers, guides) and the natural landscape.

One of the members of Yamashiro Ōboke Yōkai Village is Shōichi Shimooka. He has been gathering folklore since around the year 2000 (Shimooka, 2015). Shimooka edited the books *Yōkai mura densetsu* (Yōkai village folklore – Shimooka, 2009), *Otoroshiya* ('Frightening' in the local dialect – Shimooka, 2012) and *Omoshiro Yamashiro Tanuki-banashi* ('Interesting Yamashiro raccoon dog stories' – Shimooka, 2017), all of which were published by Yamashiro Ōboke Yōkai Village. Shimooka, also known as *yōkai hakase* (Dr Yōkai) or *yōkai jiisan* (Grandpa Yōkai), sometimes visits the Yōkai-yashiki to Ishi no Hakubutsukan (Yōkai House and Stone Museum, hereafter Yōkai-yashiki) to receive questions from children and explain yōkai culture to visitors (Figure 6.5).[8]

On 12 November 2017, I attended the 17th Yōkai Matsuri Festival at the former Kamimyō elementary school (Figure 6.6). Shimomyō elementary school students thought of ideas for yōkai sweets and sold them at the festival with cards featuring explanations of the foods and sketches of yōkai. Additionally, on the previous day, the 3rd Yamashiro Tanuki (raccoon dog) Matsuri Festival took place at Awa-kawaguchi station (Figure 6.7). An unveiling ceremony for a raccoon dog-themed station building was also held (Hitsumoto, 2017).

In all these ways, the inhabitants of Yamashiro Town pass down yōkai culture by holding exhibitions, participating in festivals and retelling local legends to younger generations. In 2013, Yamashiro Ōboke Yōkai Village was awarded the Suntory Prize for Community Cultural Activities. The local authority, Miyoshi City, cooperates with them and supports their activities.

Yōkai in the Community

Ichikawa (2013b: 2) divides community planning using yōkai culture in postwar Japan into three periods. The first period is from the 1950s to

Figure 6.5 Doctor Yōkai (Shōichi Shimooka) in Yōkai-yashiki (5 August 2018). Author's photo

Figure 6.6 Yōkai from Yamashiro walking together at the Yōkai Matsuri Festival (12 November 2017). Author's photo

Figure 6.7 Tanuki Matsuri Festival (11 November 2017). Author's photo

mid-1960s (embryonic phase), the second is from the 1970s to the beginning of the 1990s, during which local government took the initiative, and the third is from the 2000s, during which the inhabitants themselves took the initiative. Ichikawa classifies the cases of Tōno and Sakaiminato as belonging to the second period. However, the Yamashiro case is distinct from Tōno and Sakaiminato. After the rediscovery of the Konaki-jiji legend, Yamashiro developed its own original style of cooperation between inhabitants. In this section, I focus on Yamashiro. Yōkai seem to give local people a way to interpret certain phenomena, and provide a way to mediate communication among residents, between tourists and residents, and between tourists and yōkai. The former function of yōkai can be found in residents' narratives and records, while the latter can be found in residents' activities, especially concerning food.

According to Shimooka (2018), in 1995, while he was living in Fujinomiya City in Shizuoka Prefecture, he was invited to a meeting for fellow Yamashiro natives who were resident in Osaka. There, he was asked for suggestions regarding how to revitalize their hometown. Shimooka read the official town histories edited by Kondō (1960) and Tamura (1968). He found that they contained descriptions of folklore, which he found surprising because such town histories are usually only based on administrative records. He also found records of oral history and local customs, and he believes that while the vast majority of these stories and customs are not written down in documents and records, it is precisely these things that are not recorded that are important. He began to meet older people to collect their memories of stories which they had once heard. He wrote in the preface of one of the books that he edited (*Yōkai mura densetsu*):

> Yōkai in Yamashiro are necessities for coexisting with a harsh natural environment, a product of parental love wishing for children's safety [...] This is why here, we hold not just fear towards yōkai but also tender and warmhearted feelings [...] Yōkai are important treasures that we should pass on to future generations. (Shimooka, 2009: preface)

According to Shimooka, yōkai stories play a number of roles in everyday life. These include protecting children, forgiving people (by, for example, attributing their mistakes to yōkai), a way to stop dwelling on things and as warnings. In other words, they are wisdom that serves to maintain the community or a mechanism for sustaining the community (Shimooka, 2018). This wisdom comes from the danger of the mountains and fear of the darkness. Yōkai, therefore, are a frightening presence, but one that people want to see and meet (Shimooka, 2018). The Yōkai-yashiki exhibits visualized images of yōkai: handmade yōkai dolls produced by local residents, which are embodiments of the inhabitants' memories.[9] These efforts to protect and record yōkai memories have led not only to revitalization of the town but also to yōkai tourism.

Using these legends, new types of activities have also been created thorough mediums such as food. Yamashiro Ōboke Yōkai Village and the Yamashiro Tea Industry Association cooperated to develop yōkai tea (Tokushima Shinbun, 2009). Kazuyoshi Ōhira, a sous chef at Hotel Ōboke-kyō Man-naka (meaning 'right in the middle', as it is in the centre of Shikoku), made yōkai nabe (stew) with wild boar meat as a school meal for the students (Abe, 2015a). Shimomyō elementary school students invented yōkai sweets (Cha Ichigo Daifuku, a soft rice cake stuffed with green-tea-flavoured bean paste and strawberry) (Abe, 2016c). Yamashiro Shinkō Co. Ltd developed udon noodles named udon-danuki ('udon raccoon dog', which is a local Yamashiro yōkai) (Satō, 2016). Furthermore, at the Restaurant Ōboke-kyō Man-naka in November 2017, I ate Hitotsu-me don, named after the Hitotsume-nyūdō legend (regarding this legend, see Shimooka, 2009: 13–14). Communication thorough yōkai food brings together children and adults, residents and visitors.

Besides food, there are several other examples of communication between children and tourists. One example is the yōkai app, which was produced by students at Shimomyō elementary school (Abe, 2016d, 2017). If the tourists stand in front of a yōkai monument, the voices of students imitating yōkai are played, and yōkai pictures drawn by the children appear on the screen (Abe, 2017). Another example is handicrafts. On 24 November 2018, when I visited the 4th Yamashiro Tanuki Matsuri Festival, fifth-grade Shimomyō elementary school students were selling wooden coasters and wooden picture frames. At their stall, they explained to me how to use the wooden picture frames. In place of a picture, they inserted hand-drawn pictures of local Yamashiro yōkai with written messages, saying 'please display a commemorative picture of your journey'. They gave customers handmade leaflets with their own drawings of yōkai and what they learned from the story. These activities create direct interaction between children and tourists, showing how yōkai legends become a tool of communication.

International Communication Through Yōkai

Yōkai can also be a means of bringing about communication and relationships between Japan and other countries. Through yōkai, Yamashiro Ōboke Yōkai Village has built ties with Xitou Monster Village[10] in Nantou Prefecture, central Taiwan, which shows the potential to generate new ways of communicating in the present day.

Katsuyuki Ōhira is a member of Yamashiro Ōboke Yōkai Village who also runs the Ōboke-kyō Man-naka restaurant. Soon after the Great East Japan Earthquake in 2011, he received donations from Taiwanese tourists, which he delivered to the disaster areas. Ōhira and his wife were deeply impressed and strengthened their ties with Taiwan (Ōhira, 2018).[11] Ōhira became acquainted with Kuang-Yen Lin, the president of Xitou Monster Village, in 2013 (Abe, 2016a) and they worked together as friends to promote mutual exchange. These efforts were primarily on Ōhira's part, and the local government was not involved. The opening of communication with Xitou Monster Village, the similarities of yōkai culture, Japanese images in Xitou Monster Village and the regions' shared status as tea-production areas formed the basis for exchange. In August 2015, Ōhira participated in a parade wearing the costume of a local Yamashiro yōkai at the Xitou Monster Village (Abe, 2015b; Abe 2016a). On 22 November 2015, the Yōkai Matsuri festival was held in Yamashiro Town and Xitou Monster Village staff visited the festival with their yōkai costumes (Abe, 2015c). On 8 February 2016, Yamashiro Ōboke Yōkai Village and Xitou Monster Village Marketplace Development Association concluded a friendship agreement (Abe, 2016a, 2016b). Since 2015, international cultural exchange via yōkai has continued in this manner.

Xitou Monster Village opened in 2011. According to Chih-Ying Lin (2018), the general manager of Xitou Monster Village, his grandfather Shan Lin (1894–1979) was a central figure in the Village's establishment. Shan Lin was Taiwanese, but also went by the Japanese name of Shōichi Matsubayashi during the Japanese occupation of Taiwan. He used to work at a research forest belonging to Tokyo Imperial University (which since 1950 has been an experimental forest of National Taiwan University, and adjoins the Monster Village). At that time, the Japanese manager of the forest, Mr Kubota, was close in age to Shan Lin and they developed a good relationship. In 1971, Shan Lin established Mingshan Hotel (present-day Xitou Monster Village). Originally there was only forest in the area, and the only people who visited were hikers (Lin, 2018). Subsequently, Chih-Ying Lin decided to create a village based on a legend related to Xitou and his grandfather. He chose monsters as a motif for the theme park because he thought a tourist attraction needed to have a story (Yamanaka *et al.*, 2015: 47). Xitou Monster Village started as a part of Sung-Lin-Ding Shopping Street, which was built in 2009, the Monster Village following in 2011.[12]

The Monster Village has many Japanese connections. Among these is an izakaya restaurant called Rigen. During my fieldwork there in February 2018, the background music included the theme song of the famous Japanese anime film *Tonari no Totoro* (*My Neighbor Totoro*). Since 2010, there has also been a Mr Kubota bakery to commemorate the friendship between Shan Lin and Mr Kubota (Xitou Monster Village Marketplace Development Association, 2013). Inside another building, a present (a carved wooden bear) from Mr Kubota is also on display. The central torii (a gate at the entrance to a Shintō shrine in its usual Japanese context) near the entrance is to memorialize Shan Lin (Lin, 2018). Shan Lin was well liked and did many things for the community in Xitou, and as a result he is well known there. The torii is a memorial to him and his accomplishments. As Shan Lin lived through the Japanese colonial period, the torii also relates directly to that history.

Ryūhei Itō (2014: 229) uses the adjective *kawaii* ('cute') to describe the yōkai in Xitou Monster Village, interpreting Xitou Monster Village as having been influenced by Japanese yōkai culture. In a questionnaire survey conducted in 2014 by Yamanaka *et al.* (2015: 45), one answer from a tourist mentioned the village's 'kawaii Japanese atmosphere'. Yamanaka *et al.* interpreted this as showing that tourists see the yōkai as characters. The most prominent characters at Xitou Monster Village are Kumar and Badou (Figure 6.8). Kumar and Badou are animal-oriented monsters that were based on the events of a local story from the forest.[13] According to Lin (2018), Kumar and Badou were designed as *kawaii* because simple designs are easy to understand, easy to draw and easy to remember, as with other cute characters such as Doraemon and Hello Kitty.

Yamanaka *et al.* (2015: 47) note that yōkai that have become characters have transformed into something independent of the stories from the specific places and periods of time from where they originated. Masanobu Kagawa (2005: 298–299) states that in the case of kawaii yōkai, their details have been omitted, and so the yōkai are rounded off, simplified, turned into symbols and their unsettling attributes removed, becoming like a 'character'. Kagawa (2005: 299) states that the reason that modern people like *bakemono* ('monsters') which are not present in real life is that they transcend reality. Kagawa (2013: 279) also notes that the cute yōkai came to embody a kind of longing for a separate world. In this sense, Kumar and Badou are far removed from the traditional view of yōkai, and

Figure 6.8 Badou (right) and Kumar in Xitou Monster Village (25 August 2018). Author's photo

are closer to fantasy, which provides tourists with dreams and an imagined perspective.[14]

The friendly relationship between Yamashiro Ōboke Yōkai Village and Xitou Monster Village shows that yōkai have taken on a mediatory role, even though both similarities and differences exist. Both of them cherish the invisible presence that yokai represent, even as they create different types of yōkai images. They show respect for each other, and their interest in yōkai supports their friendship. Lin (2018) says that yōkai are something that bring about a common interest, even between different countries, so creating shared ground for discussion.

Conclusion

This chapter has examined several types of yōkai tourism in Japan and Taiwan. In the first section, I clarified the characteristics of three Kai heritage areas in Japan. In Sakaiminato, many yōkai from all over Japan have been collected together in the Mizuki Shigeru Road. This shows the accumulation of memories across time and place. Each yōkai has its own hometown, and originally they were spoken of only in that particular area in a particular period. Now, however, they are gathered in Sakaiminato in a way that transcends their various ages, hometowns and stories. This mixture of time and place plays an interesting role in Sakaiminato's contents tourism. In Tōno, meanwhile, storytellers convey tales of the past and *maburitto* protect that folklore over time. On the other hand, cute yōkai can also be observed here. In Yamashiro, folklore is conveyed in books, festivals and foods, and so on.

In the second section, I analysed how yōkai function in their community, focusing on Yamashiro. Shimooka (2018) considers yōkai to be a mechanism for sustaining the community, which means that yōkai are an important presence. I also focused on children's cooking and handicraft activities in Yamashiro, some of which are related to or resulted in contents tourism.

In the third section, I referred to international communication between Taiwan and Japan through yōkai. I described the process that led to the start of this communication, while showing that the two sites have a common interest in the invisible presence that yōkai represent. Kumar and Badou from the Xitou Monster Village, while based on legends from the forest, are cute monsters, leading us to fantasy.

Through the above, I have demonstrated two facets of yōkai tourism. First, yōkai tourism is of relevance to gaps of time and sometimes place. Yōkai tourism connects the past and present, and one place with another. Although the main content is the stories and concepts of yōkai themselves, so long as this foundation is present, there is the freedom to interpret them creatively. This is not possible if the original narratives and concepts are absent. For example, the fact that the story of the Konaki-jiji existed allows us to tell it, share it and enjoy it, even if we are dislocated from the specific place and period that produced it. Yōkai stories, in the minds of local people, let them imagine a period of time, a story, and its history. In this sense, the types of contents tourism discussed by Butler in Chapter 5 provide further suggestions on this topic.[15] However, note that cute yōkai (simplified characters) can sometimes exist entirely independently of the concepts they were originally based on.

Second, the relationship between popular culture and cultural heritage is important. Intangible stories turn into tangible yōkai. Invisible stories turn into visible yōkai. These yōkai can be viewed as popular culture, but if we focus on the original stories, they are intangible heritage born out of wisdom generated through daily life. While they can be treated as simple fun, yōkai also possess a deeper ethnological and anthropological meaning. One of the roles of yōkai tourism, therefore, is to invite us to think of these dual interpretations.

Yōkai in Sakaiminato are images that have been popularized for mass consumption. Because Shigeru Mizuki's manga/anime are widely known, we recognize them as familiar yōkai. In Tōno, Kappa is mainly depicted as a cute, familiar character, even though that is far from the original folklore. On the other hand, the visualized yōkai in Yamashiro are not cute. They are more complicated and realistic. They seem to embody a harsh natural environment and parental love for children, which does not allow their depiction in cute, simplified images. Kumar and Badou in

Xitou Monster Village represent the opposite extreme. They are simplified, evoking fantasy, entertainment and amusement. Accordingly, yōkai symbolization in Yamashiro and in Xitou Monster Village appear to have entirely different vectors. However, they share in common the expression of invisible yōkai as visible figures. This practice of giving invisible creatures visible figures allows us to consume their images as contents. Consequently, it produces contents tourism. As shown in this chapter, the various forms that contents tourism may take cannot be reduced to just one particular type. Rather, there are also cases (Tōno, Yamashiro) in which both the images consumed and the efforts of people to pass down stories generate contents tourism.

As seen in Tōno's case, contents tourism consists not only of the consumption of cute yōkai images (Kappa) but also the efforts of inhabitants (such as Uncle Kappa) who pass down stories. Similarly, in Yamashiro, it consists not only of the consumption of frightening yōkai images (at Yōkai Yashiki) but also the efforts of inhabitants (such as Doctor Yōkai) who pass down stories. In each case, the latter provide us with the imaginative leeway to visualize life in the past. Consumption of visualized images of yōkai (cute/non cute) can co-exist with and reinforce another imaginative world. In this sense, 'consuming images' and 'passing down stories' can be synergistic. They mutually influence each other, leading to resonance, but do not fully converge.

Notes

(1) I conducted fieldwork in Tōno (25–26 November 2016, 11–12 and 24 February 2017, 22–23 July 2018 and 20 October 2018), Sakaiminato (12–15 October 2017 and 8 July 2018), Yamashiro (10–11 October 2017, 10–12 November 2017, 4–7 August 2018 and 24–25 November 2018) and Taiwan (4–7 Feburuary 2018 and 24–27 August 2018).
(2) Within Yamashiro Town, it was Yamashiro Ōboke Yōkai Village that was certified, but the folklore concerned remains across the town as a whole (Hirata, 2018b).
(3) During my research, I also found another example of children's activity in Sakaiminato. At Kai Forum held in Yamashiro on 12 November 2017, the three Kai heritage areas were connected via broadcast. The broadcast included a video segment in which Sakai elementary school students acted as tourist guides, standing by the bronze statues of yōkai and providing explanations of them to tourists.
(4) *Mukashibanashi* ('old tales') is an oral literature tradition in Japan. When told, these stories are opened and closed with certain phrases. According to Saori Maekawa (2018) from the Tōno Culture Research Center, the phrase used to close a story in the northern Tōhoku region, the eastern part in particular, is '*Dondohare*'. There are three main interpretations of its meaning, one of which is provided by Kizen Sasaki. In 'Nōmin no bungaku', Sasaki (1987: 572–573) writes that Kyōgen (Noh comedy) plays during performances of Kagura dance are related to the most primitive form of *Mukashibanashi*, and therefore in that region several forms of Kyōgen can be observed. The Kagura performers rub a conch shell and hit a Taiko drum to ward off the devil, crying '*Dondonharai*' at the end.
(5) Shimooka (2009: preface) writes 'Kubikireuma' instead of 'Kubinashiuma'. Other sources can be found using the expression of 'Kubikire (nashi) uma' (Shimooka, 2012: 11–12). Yanagita (1938a: 12) writes 'Konaki-didi', while in the same year Akira

Takeda (1938: 10), writes 'Konaki-jiji'. Today it is generally known as 'Konaki-jijii', but throughout this chapter I adopt the spelling 'Konaki-jiji'.
(6) These three people were Gorō Hirata, Tatsuichi Hirata and Minoru Matsumoto (Takita 2002a). Gorō Hirata (1929–2000) was the father of Masahiro Hirata (the Head of Office at Yamashiro Ōboke Yōkai Village) (Hirata, 2018a).
(7) As of 2018, the village is formed of 14 full-member organizations and one observer organization (Hirata, 2019). The Director of the village is Takashi Miyamoto.
(8) Yōkai-yashiki is a fee-charging museum inside the roadside service area Michi no Eki Ōboke. It opened in 2010 and had 32,698 visitors that year (Yamashiro Shinkō Co. Ltd, 2018). I attended the sixth summer festival at Yōkai-yashiki on 4 August 2018, where the children enjoyed a haunted house. According to Kazuaki Nagamoto (2018), who is of the Representative Director of Yamashiro Shinkō Co. Ltd, which runs the *Yōkai-yashiki*, local children are accustomed to yōkai, so on this occasion, they planned a scarier haunted house than normal.
(9) Regarding the function of Yōkai-yashiki and the revival of folklore space, see Ichikawa (2016b).
(10) In the village, a string of three Chinese characters 妖怪村 (Yaoguai Cun in pinyin) can be seen in many places in the village. These are the same characters read as 'yōkai mura [village]' in Japanese.
(11) This is the reason that Ōhira accepts many Taiwanese students for internships at Ōboke-kyō Man-naka Group (operating a restaurant and a hotel).
(12) The main designer at the village is Jiun-Lin Tzeng, and Xitou Monster Village is full of ideas, merchandise and characters that he has created.
(13) According to the legend, Matsubayashi [Shan Lin] owned two dogs, a small bear named Kumar and a small leopard named Badou. One day, he was attacked by a monster. Kumar and Badou helped him, but Badou was injured and passed away, and Kumar went missing. Kumar was originally a Formosan black bear and Badou was originally a Formosan clouded leopard, but in the story, they have been turned into monsters (Lo, 2019).
(14) Other monsters include A-Sa (with a green face) and Blue (with a red face). According to legend, Tengu (a kind of yōkai that lives deep in the mountains in its usual Japanese context) is the guardian of the Monster Village and this area of Xitou. A-Sa and Blue, considered to be mountain monsters, are followers of Tengu. A-Sa controls people's academic performance and health, while Blue presides over careers, love and relationships. In Blue's world, everything appears blue, hence the character's name (Lo, 2019). A-Sa and Blue resemble Japanese *namahage*, so maintain a traditional Japanese yōkai image. In this way, kawaii yōkai and traditional Japanese-like yōkai co-exist.
(15) Ichikawa (2016a, 2017) and Foster (2015: 95–96) also give examples of imaginative creativity regarding yōkai images.

References

Abe, K. (2015a) 'Yōkai nabe' ni jidō ōyorokobi. *Tokushima Shinbun*, 15 May, p. 14.
Abe, K. (2015b) Yōkai de kokusai kōryū, Taiwan kara kyara hatsu sansen. *Tokushima Shinbun*, 28 October, p. 17.
Abe, K. (2015c) Taiwan no Yōkai to fureau. *Tokushima Shinbun*, 23 November, p. 20.
Abe, K. (2016a) Taiwan kyōkai to kyōtei. *Tokushima Shinbun*, 6 February, p. 20.
Abe, K. (2016b) Yōkai no miryoku bankoku kyōtsū. *Tokushima Shinbun*, 9 February, p. 12.
Abe, K. (2016c) Yōkai suiitsu ikaga? *Tokushima Shinbun*, 12 February, p. 10.
Abe, K. (2016d) Yōkai shōkai apuri kaihatsu. *Tokushima Shinbun*, 27 December, p. 17.
Abe, K. (2017) Yōkai shōkai apuri kansei. *Tokushima Shinbun*, 20 January, p. 12.

Denshōen (2010) Denshōen Homepage. See www.densyoen.jp (accessed 10 December 2018).

Foster, M.D. (2015) *The Book of Yōkai: Mysterious Creatures of Japanese Folklore*. Oakland, CA: University of California Press.

Hirata, M. (2018a) Conversations between the author and Head of Office at Yamashiro Ōboke Yōkai Village during a tour of historical sites in Yamashiro Town, 6–7 August.

Hirata, M. (2018b) Interview conducted by the author via email, 26 September.

Hirata, M. (2019) Interview conducted by the author via email, 18 January.

Hitsumoto, M. (2017) 'Kisha-danuki' ekisha ohirome. *Tokushima Shinbun*, 12 November, p. 17.

Ichikawa, H. (2013a) Chiiki shakai ni okeru yōkai kan no keisei to keishō. *Bunka Shigen Gaku* 11, 127–138.

Ichikawa, H. (2013b) Machizukuri ni okeru yōkai bunka no gendaiteki katsuyō ni kansuru kōsatsu. *Geijutsugaku kenkyū* 18, 1–10.

Ichikawa, H. (2016a) Yōkai saishū no susume: nichijō wo kakuchō suru manazashi no kakutoku ni mukete. In H. Imai and H. Ōmichi (eds) *Kai'i wo Aruku: Kai'i no Jikū 1* (pp. 89–111). Tokyo: Seikyūsha.

Ichikawa, H. (2016b) Michi no eki wo kiten to suru eco museum teki jissen wo tōsita denshō kūkan no saisei: Shikoku no hikyō Yamashiro Ōboke yōkai mura no torikumi kara. *Ecomuseum Kenkyū* 21, 30–39.

Ichikawa, H. (2017) Tsukurareru yōkai tachi: Chiiki ni nezashita monogatari saisei e no kokoromi. In K. Komatsu (ed.) *Shinka suru Yōkai Bunka Kenkyū* (pp. 290–306). Tokyo: Serika Shobō.

Itō, R. (2014) Taiwan yōkai mura tanbō ki. In *Kai* 41 (pp. 226–229). Tokyo: Kadokawa Shoten.

Kagawa, M. (2005) *Edo no Yōkai Kakumei*. Tokyo: Kawade Shobō Shinsha.

Kagawa, M. (2013) *Edo no Yōkai Kakumei* (revised edition from Kawade Shobō Shinsha in 2005). Tokyo: Kadokawa Gakugei Shuppan.

Komatsu, K. (2003) Yōkai to yōkai kenkyū: Joron ni kaete. In K. Komatsu (ed.) *Nihon Yōkaigaku Taizen* (pp. 9–28). Tokyo: Shōgakukan.

Komatsu, K. (2006) *Yōkai Bunka Nyūmon*. Tokyo: Serika Shobō.

Komatsu, K. (2007) *Yōkai-gaku Shinkō*. Tokyo: Yōsensha.

Kondō, T. (ed.) (1960) *Yamashirodani Sonshi*. Yamashiro: Yamashiro Town Office (Miyoshi District, Tokushima Prefecture).

Lin, C.Y. (2018) Interview conducted by the author at Xitou Monster Village, Taiwan. 25 August.

Lo, Y.C. (2019) Interview conducted by the author via email, 20 January.

Maekawa, S. (2018) Interview conducted by the author at Tōno Culture Research Center in Tōno, 23 July and interview via email, 12 November.

Mizuki Production and Imamura, K. (eds) (2013) *GeGeGe no Kitarō no Yōkai to Jōzu ni Tsukiaō!!* Sakaiminato: Mizuki Shigeru Museum.

Nagamoto, K. (2018) Interview conducted by the author at Yōkai-yashiki in Yamashiro, 5 August.

Ōhira, K. (2018) Interview conducted by the author at Hotel Ōboke-kyō Man-naka in Yamashiro, 5 August.

Sakaiminato City (2017) Sakaiminato City Statistics on Mizuki Shigeru Road and Mizuki Shigeru Museum (received at Mizuki Shigeru Museum on 12 October 2017).

Sakaiminato City Tourism Association (2018) Mizuki Shigeru Rōdo no ayumi (received from Sakaiminato City Tourism Association via post, 15 November 2018).

Sasaki, K. (1987) Nōmin no bungaku. In Tōno Municipal Museum (ed.) *Sasaki Kizen Zensyū II* (pp. 571–576). Tōno: Tōno Municipal Museum (originally published 1930).

Satō, Y. (2016) Deta? Yōkai 'udon-danuki'. *Tokushima Shinbun*, 8 May, p. 17.

Shimooka, S. (ed.) (2009) *Konaki-jiji no Sato Yōkai Mura Densetsu*. Miyoshi: Shikoku-no-hikyō Yamashiro Ōboke Yōkai Mura.

Shimooka, S. (ed.) (2012) *Yōkai Mura Densetsu Otoroshiya*. Miyoshi: Shikoku-no-hikyō Yamashiro Ōboke Yōkai Mura.

Shimooka, S. (2015) Furusato ni wa takaramono ga ippai, yōkai densetsu hitobito wo miryō. *Tokushima Shinbun*, 11 September, p. 10.

Shimooka, S. (ed.) (2017) *Omoshiro Yamashiro Tanuki Banashi*. Miyoshi: Shikoku-no-hikyō Yamashiro Ōboke Yōkai Mura.

Shimooka, S. (2018) Interview conducted by the author at Yōkai-yashiki in Yamashiro, 5 August.

Shōji, Y. (2018) Interview conducted by the author at the Mizuki Shigeru Museum in Sakaiminato, 8 July.

Takeda, A. (1938) Sanson Goi. *Minkan Denshō* 4 (2) (Minkan Denshō no Kai), 10.

Takita, M. (2002a) Yanagita Kunio no yōkai 'Konaki-jiji' zō to hi. Unpublished.

Takita, M. (2002b) 'GeGeGe no Kitarō' no mangaka Mizuki Shigeru shi wo tazunete. *Tokushima Shinbun*, 4 July, p. 12.

Tamura, T. (ed.) (1968) *Sanmyō Sonshi*. Yamashiro: Yamashiro Town Office (Miyoshi District, Tokushima Prefecture).

Tokushima Shinbun (1999) Ken-nai no san-kan bu de denshō kakunin. *Tokushima Shinbun*, 3 September, p. 13.

Tokushima Shinbun (2009) 'Yōkai-cha' ikaga? *Tokushima Shinbun*, 22 April, p. 27.

Unman, H. (2018) Interview conducted by the author at Denshōen in Tōno, 22 July.

Xitou Monster Village Marketplace Development Association (2013) *Monster News*, January (newsletter in Taiwanese).

Yamamura, T. (2011) *Anime, Manga de Chiiki Shinkō: Machi no Fan wo Umu Kontentsu Tsurizumu Kaihatsuhō*. Tokyo: Tokyo Hōrei Shuppan.

Yamanaka, C., Itō, Y. and Momose, H. (2015) Popyurā bunka no kankō shigen ka to 'dentō no sōzō: Taiwan Nantou ken Siitō yōkai mura wo jirei to shite. *Jin-ai Daigaku Kenkyu Kiyō Ningen Gakubu Hen* 14, 39–50.

Yamashiro Ōboke Yōkai Mura (2008) *Yōkai Mura Zukuri Keikakusho*. Miyoshi: Shikoku-no-hikyō Yamashiro Ōboke Yōkai Mura.

Yamashiro Shinkō Co. Ltd (2018) Visitor data for Michi no Eki Ōboke and Yōkai-yashiki to Ishi no Hakubutsukan (received from Yamashiro Shinko Co. Ltd on 11 May 2018 via email).

Yanagita, K. (1938a) Yōkai meii. *Minkan Denshō* 3 (10) (Minkan Denshō no Kai), 12.

Yanagita, K. (1938b) Yōkai meii (4). *Minkan Denshō* 4 (1) (Minkan Denshō no Kai), 12.

Yanagita, K. (1939) Yōkai meii (6). *Minkan Denshō* 4 (6) (Minkan Denshō no Kai), 16.

7 Contents Tourism and Religious Imagination

Kyungjae Jang

> It is an obligation. Although I'm a fan of other characters, to climb
> this mountain is my duty as a Love Liver [fan of *Love Live!*]
> Interviewee U-3, March 2017

This chapter clarifies religious aspects of contents tourism through a study of Korean fans' pilgrimages relating to the Japanese mixed-media contents *Love Live! School Idol Project*. In the introduction of this book, Yamamura suggested the concept of contentsization, which refers to 'the continual process of the development and expansion of the "narrative world" through both mediatized adaptation and tourism practice'. One such type of expansion that occurs during contentsization is the addition of a sacred meaning to the places related to contents. Visiting a location related to the contents is usually called pilgrimage. Fans as pilgrims often visit a particular place several times, leaving traces such as notes at the site. Other more active fans perform more religious-looking performances. The examples given in this chapter include fans who bow and make altar-like constructions at the sites they visit. I discuss if these actions are actually religious, or whether they are simply taking on the form of an existing rite to attract attention from others, particularly users of social media. The example of *Love Live*'s Korean fans allows us to explore the nature of the 'religious imagination' within contents tourism.

Popular Culture Fandoms and Religion

Fans of popular culture show devotion and loyalty, in ways often similar in form to religious beliefs and devotions, to their favourite contents. Given these similarities, many researchers have studied religion in relation to popular culture fandom (Davidsen, 2013; Giles, 2000; Possamai, 2005; Rojek, 2001). Giles (2000: 135) argues that the Star Trek fandom is like a kind of civil religion in that it has an organization, a dogma and a recruitment system, and that the same phenomena underpinned religion in the Middle Ages. The Star Trek fandom is defined as a 'new cultural religion'.

Meanwhile, Possamai's (2005) study of Jedi Religion defines it as 'hyper-real religion'. Davidsen (2013) suggests the term 'fiction-based religion' and describes instances of religious activity in which fictional texts are used as authoritative texts. What is common to these studies is that similarities with religion appear in the fandom, and for that reason the fandom is seen as a religion.

On the other hand, not all researchers agree that fandoms may be seen as religions. Many researchers have said that pop culture is not the same as religion or a substitute for religion just because comparisons can be made (Cavicchi, 1998; Duffett, 2013; Hills, 2002). Cavicchi criticizes the equation of fandom with religion as follows:

> Ted Harrison wrote a book on Elvis fans in which he argued that Elvis fan culture represents 'a religion in embryo'. In particular, he outlined the religious aspects of the culture, such as an emphasis on an 'Elvis Gospel', the worship of 'icons and relics', and the power of the Elvis 'resurrection myth'. (Harrison, 1992)

> Such interpretations miss the point. Fans' use of religious language in explaining and thinking about fandom and the clear parallels between their behavior and that of Christian believers do not mean that fandom is a religion; rather, they point to the fact that both fandom and religion are addressing similar concerns and engaging people in similar ways. (Cavicchi, 1998: 187)

Furthermore, Duffett (2013: 146) identifies the reasons why it is difficult to see fandom as a substitute for religion: 'These are, first, that it has no central theology, second that fans can "worship" more than one "deity" at a time, and third, that the idea rests on a questionable conception of human need'.

So, some researchers have said fandoms display religiousness rather than actually being religions. The concept of 'neo-religiosity' proposed by Matt Hills (2002) is representative. However, Duffett (2013) criticizes such a view of fandom as a religion as conceptual, and says that the idea of a religion based on fandom has not yet been firmly established. Consequently, there is a wide range of discussion regarding whether a fandom may be called a religion or whether fandom substitutes for religion. Hills (2002: 118) stated that fandoms exist between cult and culture, whether or not they are similar to religion, and linking fan culture to religion or religiousness is a difficult issue for both fans and scholars owing to the difficulty of defining what a religion actually is.

Religious Meanings of Pilgrimage in Contents Tourism

Pop culture fans visit places related to their favourite works. The works can be in any media format, including movies, dramas, literature, music and animation, and the phenomenon is global and can be seen on every continent. Fans do not simply visit places. They also add meaning to the media-related places via their travel. In many instances, the fans themselves call their visits 'pilgrimages', and the term has become widely used by scholars, too. The word pilgrimage is also being used in a lighter sense than in the past. Bowman (2008: 279) states:

> modern Pilgrims visiting the sacred places are of many different beliefs and often of no belief. They do not necessarily come to be in touch with any specific divinity but they come to be in the energies of the sacred places and by being in these places to understand themselves more clearly and to see their role in the world.

In other words, the sites of pilgrimage may be determined originally from religious beliefs, but any site visitors think has special meaning can become a sacred place. As Margry writes, 'Any place where people met occasionally or en masse to pay their respects to a special deceased person soon came to be referred to as a "place of pilgrimage", although it was not clear what this actually meant' (Margry, 2008: 17).

The term 'pilgrimage' to mean travel motivated by works of popular culture was coined by fans. Fans often use religious terminology for their fannish activities. For example, Cavicchi (1998: 95) notes, 'In fact, in interviews, fans often referred to a Springsteen performance as a "religious experience" or "spiritual experience."' Such vernacular has affected researchers' discussions of religiosity and popular culture fandoms. So why did the fans use religious terminology to describe their travel? Duffett (2013: 145) says that 'religious language could be no more than a convenience'. In his analysis of the studio tour of the soap opera *Coronation Street* in the UK, Couldry said, 'The set is not any space, any street, but the actual street that you and everyone else have been watching from your home. It is, in this precise sense, a ritual place, where two "worlds" are connected' (Couldry, 2005: 71).

This connection is a symbolic interaction, not simply a link between a place that appears in the media and an actual place. Aden (1999: 10) refers to pop culture fans' 'symbolic pilgrimages' as 'the interaction of story and individual imagination', in other words, a link between an individual, the contents, and a place. Furthermore, 'Rather than embarking upon a "real life" journey to a holy shrine, symbolic pilgrimages feature individuals ritualistically revising powerful places that are symbolically envisioned through the interaction of the story and individual imagination' (Aden, 1999: 10). Importantly, the experience of these individuals is shared and reproduced in the community of the fandom. 'Being a fan is being a member of a purposeful play community in which individual interpretations remain unique yet overlap with others' interpretations' (Aden, 1999: 10).

The internet, especially social media, plays an important role in pop-culture-related pilgrimage. The internet and social media have roles in providing information on 'must-visit' places (Xiang & Gretzel, 2010), marketing tourism sites (Hays *et al.*, 2013) and sharing tourism experiences (Munar & Jacobsen, 2014). Jansson (2018) emphasizes the role of photos uploaded to social media sites:

> Tourists, like people in general, are to an increasing extent (co-)producers of media texts that can be spread and discussed far beyond the close circles of traditional family albums. Instagram images, for example, can be geotagged and immediately commented upon, which in turn contributes to the cultural (re)coding of tourism places and practices. (Jansson, 2018: 102)

Such photographs provide evidence of the authenticity of space:

> The boundaries between the fictional world of the media content and the existing landscape seem to vanish into what Urry and Larsen (2011) call the 'mediatised gaze' (p. 115). By posting their tourist experiences on social media, the fans' pilgrimage photos might coincidently become evidence for the authenticity of the local places appearing in a film production. (Scherer & Thelen, 2018: 80)

This process also works in the opposite direction, whereby photos taken and those photographed acquire authenticity in the social media space. These activities often occur in pop culture fandoms when fans visit places related to their favourite works and make religious-looking poses. Photos of religious-looking poses are shared via social media, and sometimes convey the meaning that the site where the photos were taken has religious-like meaning. One Korean internet community user named Taiga (2014) collected religious-looking pilgrimage photos of *Love Live* fans on social media and wrote '*Love Live* is Religion. *Loblahu Akbar*' in a play on the Muslim expression *Allahu Akbar* (God is the greatest). Many other articles or posts of a similar nature can be found on various social media platforms such as Twitter.

Such actions of fans are self-projections that can, by accident or design, become quasi-canonical rituals. Consequently, it cannot be said that there is no religiosity in pop culture-related rituals. Fundamentally, people pursue those things on which they rely or for which they have special need. Pop culture plays a role in shifting the objects of worship from divine beings to people. Thus, pop culture does not only act like a religion as an effective means of self-expression, as McCloud (2003) has argued, but pop-culture-based rituals also play a role as an invented new type of religion (Bickerdike, 2016; Davidsen, 2013). To give an example of this process, the next section examines how fans of the Japanese mixed-media contents *Love Live! School Idol Project* construct their sacred places.

Love Live! as Sacred Contents for Korean Fans

The *Love Live! School Idol Project* (hereafter *Love Live!*) started in June 2010 in Japan. *Love Live!* is a multimedia project focusing on a fictional idol group. It spread to Asian countries from 2013, and was first imported to South Korea in 2014. The narratives and songs are disseminated via comic magazines, animation, games, novels and other formats. Japanese publisher ASCII Media Works (Kadokawa Corporation), the music label Lantis Company and the animation production company Sunrise jointly invested in the project. The first season, which focused on the idol group μ's (muse), officially ended in April 2016, and the second series, *Love Live Sunshine*, with the new fictional idol group Aqous, is currently underway (in late 2018) with a different set of characters and voice actresses. This chapter focuses on the first season of *Love Live!*

Love Live! is a simple story in which nine high school girls form a school idol team called μ's (muse) and promote their high school in an attempt to prevent it from closing. One of the characteristics of *Love Live!* is fans' participation in the on-going production of the franchise, based on its catchphrase *Minna de kanaeru monogatari* ('A story realized through everyone's participation'). For example, *Dengeki G's Magazine*, the monthly magazine featuring *bishōjo* (cute young girl) anime and games that originally carried the *Love Live!* story, encouraged readers to engage with the contents and even to contribute to making them. The most representative example is the unit and position decision election. In the same way that fans of the human idol group AKB48 are asked to vote for their favourite members, thereby determining their positions within dance routines and as lead singers, fans were encouraged to vote for the positions of singers within the virtual idol group μ's. Furthermore, in the 30 July 2014 edition of the magazine, a new song production plan for readers was launched. In this project, fans helped to produce a new song by voting on ideas for the lyrics or other components suggested by the readers. Voting also took place regarding costumes and choreography, and a new single was released based on the selected items. This process provided reality to the virtual idols and encouraged the devotion of fans.

Religious-like behaviours of *Love Live!* fans

One feature of the behaviour of *Love Live!* fans is that there are many religious-like and ritual-like performances. Such performances are mainly observed in East Asian countries/regions, including China, Taiwan, Singapore and South Korea. There are four ritual-like fan performances: 'call and pen light', 'armament', 'bowing' and 'birthday pilgrimage'.

First, 'call and pen light' is a part of concert culture. 'Call' (*kōru* in Japanese) is Japanized English and refers to shouting and cheering to a song, whereas *pen raito* (pen light) refers to shaking a shiny fluorescent stick in rhythm with a song's beat. The terms 'call' and 'pen light' have been adopted in other Asian countries, so in Korean, for example, the equivalents are *kol* and *pen laitu*. Like a religious ceremony, the same procedure for the call and pen light rituals has spread all over East Asia. Once the practice was created, the call and pen light ritual diffused through social media. One of the significant ritual-like forms of 'call and pen light' can be seen when performances are in movie theatres. Call and pen light started as a way for fans to show loyalty to the singers at concerts. However, in the case of movie theatres, there are only screens in front of the fans. In Korean, this is called *kol-jang-pan*, which means a movie at which fans can perform a call. Call and pen light is prohibited at general screenings, so fans pool their resources to rent an entire theatre, or a production company organizes a special screening at which fans can perform 'call and pen light' rituals (see also Sugawa-Shimada, Chapter 8).

Second, armament is a unique behaviour to *Love Live!* fans. Armament is the act of wearing a large and often chaotic array of *Love Live!* merchandise, such as badges and soft toys, when fans go to their sacred places such as Akihabara in Tokyo or comic events (Figure 7.1). Armament is practised in Japan, South Korea, Taiwan and Hong Kong. It is called *busō* in Japan, *mujang* in Korea and *quán fù wǔ zhuāng* in Chinese, and all three terms have same meaning of 'armament' in English. Armament can be distinguished from usual cosplay practices or the carrying/wearing of merchandise because it involves large decorations that go beyond the category of general costumes (see Yamamura, Introduction; Rastati, Chapter 9). Armament usually covers the whole body, but the largest armaments can be about 2 metres in height and 3 metres in length, and there are badges, tapestries, soft dolls and game cards attached all over the wearer.

Figure 7.1 Fans in armament at Seoul Comic World. Author's photo

There are fans who wear items on both the front and back of their bodies. In many cases, walking the streets and participating in events in armament can be challenging given the size and weight of the outfit.

The third behaviour of *Love Live!* fans that is similar to religious ritual but creates a contrast with other fandoms is bowing. Some fans perform kowtow (kneeling and bowing very low) to the characters at related places. Interestingly, this is the only aspect of *Love Live!*-related fan rituals that did not originate in Japan, but started in Singapore and South Korea. Bowing as a sign of respect is seen in many cultures, including Islam, Buddhist and Confucian-based East Asian societies. In Imperial China, the kowtow was performed when showing respect to the emperor. Called *jeol* in Korea, bowing while kneeling is still used as an expression of respect for a deceased person or an elder.

Love Live! fans began to perform a kowtow to show respect towards their animated idols from around March 2014. A photo of fans bowing at *Love Live!* events in Singapore and Hong Kong spread to China, Japan, South Korea and Taiwan via social media. Meanwhile, in Seoul, an advertising panel for the *Love Live!* game was installed in Hongdaeipgu metro station on 10 June 2014, and a photo of fans bowing was posted two days later (Manggazip, 2014). In China, a scene of fans bowing to a train with *Love Live!* advertising, which started running in Shanghai on 26 July 2014, was shown on the Japanese news (Sawai, 2014). Another example of sacred bowing was a photo of fans bowing in front of a *Love Live!* panel in Korean theatres, which was uploaded to the fan community on 27 September 2015, coinciding with Chuseok (Lubpokcase, 2015), a traditional Korean harvest festival. Koreans perform ancestral memorial rites, including kowtowing, on the morning of Chuseok. The act of bowing on the day of Chuseok can be interpreted as evidence that fan behaviours towards the characters of *Love Live!* intentionally create a religious image.

The fourth sacred fan behaviour is the birthday pilgrimage. Some *Love Live!* fans climb Mount Umi in South Korea because the mountain has the same name as one of the characters, Umi. Fans from all over Korea have climbed the mountain each year since 2015 on her birthday, 15 March. Actually, there are two Mount Umis in Korea, in Daegu and Goheung. The mountain in Daegu is more famous for *Love Live!* fans because it is easily accessible from many parts of South Korea, but some fans prefer Mount Umi in Goheung because the sea (*umi* in Japanese) is visible from the top of the mountain. The reason why fans climb the mountains to celebrate the character's birthday is related to the storyline in *Love Live!* In the second episode of the *Love Live!* TV animation series, which was broadcast in April 2014, Umi reveals that she likes mountains. This character detail attracted fans' attention and in October 2014 Korean fans realized that there are two mountains called Umi in Korea. In the storyline, Umi's birthday is 15 March. So Umi fans from all over Korea, including Seoul, Daegu, Busan and Gangwon-do, started to climb Mount

Umi in 2015 (Mette, 2016). Fans made a kind of altar at the top of the mountain using Umi merchandise, such as plush dolls, and kowtow in front of the altar. Fans uploaded photos of such performances on social media to promoted the mountaintop as a sacred place related to Umi.

Climbing Mount Umi Mountain as a symbolic pilgrimage

On 15 March 2017 I climbed Mount Umi in Daegu to undertake participant observation of the third year's climb by fans. *Love Live! Sunshine* had started in 2016, and previous members from the first series of *Love Live!* were replaced by new characters. I assumed that there would be no more fans visiting Mount Umi and that the fan pilgrimage there would have ended. However, as in previous years, many fans climbed Mount Umi. In total, more than 20 fans visited Mount Umi to celebrate Umi's birthday. I met eight fans on Mount Umi, and posts on social media indicated that many fans visited at the weekend. In this section I paint a portrait of fans' behaviour at the top of the mountain based on my observations and interviews with seven pilgrims who made the climb.

Mount Umi is about 800 metres high and it takes about an hour to reach the summit. The path is fairly steep, so some fans who are not accustomed to hiking call it a 'real pilgrimage'. I arrived at the bottom of the mountain at 10:50 a.m. having arrived there by bus from Daegu. I saw some people who looked like *Love Live!* pilgrims, but they did not talk to each other. I arrived at the top of the mountain at noon at around the same time as the other fans I had seen at the bottom of the mountain. It was notable how the social atmosphere had changed in the time between arrival at the bottom of the mountain and reaching the summit. As the pilgrims realized the others were also pilgrims, they greeted each other and started to converse.

U-1 (twenties, male) came from Gumi and was a fan of Maki, another character in *Love Live!* In our first conversation, he said that he did not speak to me because he did not realize I was a pilgrim. As with other fandoms, fans feel like they are in the same community and he opened up when he knew I was there for the same reason. He told me about the reasons for his visit, which was his first. He did not have a job, and called himself an '*otaku*' (the Japanese term for nerd or fan) of *Love Live!* He came because it was Umi's birthday, and he hated the new series *Love Live! Sunshine* with the new characters.

After a few minutes, U-2 (twenties, male) from Kimhae arrived at the summit. He was a fan of Umi and said that he did not speak to us at the bottom because he did not realize we were pilgrims. He started making an altar. He laid down a newspaper and put some makgeolli (a Korean rice wine) there in a manner similar to *Gosire*, which is a shamanistic ceremony performed for the spirits of one's ancestors in front of their grave (Figure 7.2).

Figure 7.2 A ritual resembling *Gosire* performed by a fan at the top of Mount Umi. Author's photo

Then, three more fans arrived. U-3 (twenties, male), U-4 (twenties, male) and U-5 (twenties, male) were from Gangwon-do, Chungbuk and Busan respectively, but had come to Mount Umi together after getting in contact with each other online. As students returning from their compulsory military service, they said that they were worried about going on a school trip the next week because they had almost no friends at the university. This is common in Korean universities. Usually, students returning from military service are called *Bok hak saeng* (return students), and they become outsiders in campus life because their friends have moved on while they were away in the military. In many cases (but not all), such students immerse themselves in games and animation. U-4 placed a doll of Kotori, another character from *Love Live!*, on the altar that U-2 had made. It was not an Umi doll, but nobody complained. U-4 said that he also had *Love Live! Sunshine* goods, but he did not take them out because U-1, who hates the *Sunshine* series, said in jest that he should take them out only if he was willing to jump off the mountain with them. The altar was completed at about 12:15 p.m., with the dolls, books and cards taken from each of the participants. U-2 sprinkled a glass of rice wine on the mountain according to the rituals of *Gosire*.

Another fan, U-6 (twenties, male), arrived at 12:48 p.m. He brought dolls of three *Love Live!* characters and put them on the altar. During the subsequent discussions with the fans, the following were topics of conversation. Umi's birthday was on a weekday, so few fans were visiting today (the actual birthday), but the pilgrims had seen posts on Twitter indicating that there were many fans who had climbed the mountain the previous weekend. The pilgrims also said that many fans visited the other Mount Umi in Gohung this year because the sea (*umi* in Japanese) is visible from the top of the mountain. Many fans like such jokes because they know some Japanese. At around 13:00, U-2, U-3, U-4 and U-5 were ready to go down. We agreed to meet again at this place every year, and U-1 and U-3 gave Umi goods to U-2, the only Umi fan in the group. Around 1:20 p.m., U-7 (twenties, male) and U-8 (twenties, male) arrived. They came together, but said that they met in person for the first time that day. U-7 was set to join the military in six days, and U-8 was going to graduate school in Tokyo the following April. Both brought Umi merchandise and they made a new altar. However, they did not do any ritual-like performance. I came down the mountain after watching them make their altar.

Conclusion

Through the example of the behaviour of Korean fans of *Love Live!* this chapter has shown how they create, maintain and propagate the value of a sacred place related to their favourite contents. It was a curious experience to see fans from all over the country climb a mountain in the countryside that has nothing to do with *Love Live!* except having a name that is the same as an anime character. The pilgrims I met at the top of Mount Umi said that mountain climbing is not their hobby. Furthermore, some of them rarely leave their houses. However, they climbed the mountain and go every year. It seems ascetic, but the fans enjoyed enduring the tough climb to the top of the mountain. Interestingly, most of the hikers were not fans of Umi, the character which inspired the act of climbing that particular mountain. However, they often said it was their obligation as a fan to perform such rituals. This creates the impression of a religious ritual. There is nothing special at the top of the mountain. It is not a location of the *Love Live!* contents, nor is there any event, nor a monument however small indicating some kind of connection with Umi. Even so, fans climb the mountain.

The unique feature of fan pilgrimage at Mount Umi is that no element other than the mountain's name is related to the contents. Other fandom activities labelled pilgrimages are to places where visitors can have specific experiences, such as standing at the location where a story was set/filmed or being close to related person such as actors and authors. Of course, there are places that are not related to the contents, but they are mostly places where there is at least a connection with the work, such as a similar atmosphere. For Umi and Mount Umi, the connection between contents and place is almost completely contingent. The symbolic representation of the mountain's name and the beliefs of those who visit it are the motives for fans to devote themselves to climbing the mountain every year.

In that sense, the case study in this chapter provides an example of the dedication, duty and faith of the fandom in contrast to the other pilgrimages of contents tourism. Of course, it is difficult to call the practices described in this chapter 'religion'. However, during the fans' visits, there were strong associations with recognizable religious rituals: hoping for good things to happen, prayers before embarking on a major endeavour (such as studying abroad), making an altar and performing offering at it. Furthermore, these were visits made without any obvious purpose beyond reverence for the contents, and some of the fans talked almost as if it was a religious duty.

In this process, social media supports the pilgrimage of *Love Live!* fans. First, social media enables fans to communicate and create communities. Next, social media plays a role in spreading the authenticity of Umi. Through social media, a message is circulated that fans should visit Mount Umi to celebrate the *Love Live!* character Umi on 15 March every year. Such a message binds the contents tourists/pilgrims together almost in the manner of a scripture. The example of climbing Mount Umi as an act of devotion to Umi, and other acts of devotion by the *Love Live!* fandom, comprise examples of religiosity, or a similarity with actual religious practices, akin to 'symbolic pilgrimage' in contents tourism.

References

Aden, R.C. (1999) *Popular Stories and Promised Lands: Fan Cultures and Symbolic Pilgrimages*. Tuscaloosa, AL: The University of Alabama Press.
Bickerdike, J.O. (2016) *The Secular Religion of Fandom: Pop Culture Pilgrim*. London: Sage.
Bowman, M. (2008) Going with the flow: Contemporary pilgrimage in Glastonbury. In P.J. Margry (ed.) *Shrines and Pilgrimage in the Modern World: New Itineraries into the Sacred* (pp. 241–280). Amsterdam: Amsterdam University Press.
Cavicchi, D. (1998) *Tramps Like Us: Music and Meaning among Springsteen Fans*. Oxford: Oxford University Press.
Couldry, N. (2005) On the actual street. In D. Crouch, R. Jackson and F. Thompson (eds) *The Media and the Tourist Imagination: Converging Cultures* (pp. 60–75). London: Routledge.
Davidsen, M.A. (2013) Fiction-based religion: Conceptualising a new category against history-based religion and fandom. *Culture and Religion* 14 (4), 378–395.
Duffett, M. (2013) *Understanding Fandom: An Introduction to the Study of Media Fan Culture*. New York, NY: Bloomsbury.
Giles, D. (2000) *Illusions of Immortality: A Psychology of Fame and Celebrity*. Basingstoke: Hampshire.
Harrison, T. (1992) *Elvis People: The Cult of the King*. London: Fount.
Hays, S., Page, S.J. and Buhalis, D. (2013) Social media as a destination marketing tool: Its use by national tourism organisations. *Current Issues in Tourism* 16 (3), 211–239.
Hills, M. (2002) *Fan Cultures*. London: Routledge.

Jansson, A. (2018) Rethinking post-tourism in the age of social media. *Annals of Tourism Research* 69, 101–110.
Lubpokcase (2015) Oneul dongdaemun pannel e jeolhangeo injeung. See http://gall.dcinside.com/lovelive/3790458/ (accessed June 2018).
Manggazip (2014) Ilweb e Lovelive! hongdaeipguyok hangukin keunjeol (dogeza) sajin hwaje, daeman, singapore loveliver sajin chuga. See http://manggazip.tistory.com/1228/ (accessed July 2018).
Margry, P.J. (2008) Secular pilgrimage: A contradiction in terms? In P.J. Margry (ed.) *Shrines and Pilgrimage in the Modern World: New Itineraries into the Sacred* (pp. 13–46). Amsterdam: Amsterdam University Press.
McCloud, S. (2003) Popular culture fandoms, the boundaries of religious studies, and the project of the self. *Culture and Religion* 4 (2), 187–206.
Mette (2016) 3wol 15il daegu dalseong-gun umisan deungjeong-gi. See http://bbs.ruliweb.com/hobby/board/300100/read/29324134/ (accessed July 2018).
Munar, A.M. and Jacobsen, J.K.S. (2014) Motivations for sharing tourism experiences through social media. *Tourism Management* 43, 46–54.
Possamai A. (2005) *Religion and Popular Culture: A Hyper-real Testament*. Brussels: Peter Lang.
Rojek, C. (2001) *Celebrity*. London: Reaktion Books.
Sawai, M. (2014). Chūgoku de 'Rabu Raibu!' no ita ressha tōjō → kangekishi dogeza suru fan ga zokushutsu shi butsugi wo kamosu/gakusha 'Ue no sedai wa bosatsu wo, wakamono wa anikyara wo ogamu. Mondai wa nai'. *Rocketnews*, 4 August. See http://rocketnews24.com/2014/08/04/473085/ (accessed July 2018).
Scherer, E. and Thelen, T. (2018) Drama off-screen: A multi-stakeholder perspective on film tourism in relation to the Japanese morning drama (asadora). In S. Kim and S. Reijnders (eds) *Film Tourism in Asia: Perspectives on Asian Tourism* (pp. 69–86). Singapore: Springer.
Taiga. (2014). Leobeulaibeuneun jong-gyoibnida. Leoblahu akeubaleu! *Ilganbeseuteu jeojangso*, 31 May. See http://www.ilbe.com/3626673324 (accessed January 2019).
Urry, J. and Larsen, J. (2011) *The Tourist Gaza 3.0*. London: Sage.
Xiang, Z. and Gretzel, U. (2010) Role of social media in online travel information search. *Tourism Management* 31 (2), 179–188.

8 The 2.5-Dimensional Theatre as a Communication Site: Non-site-specific Theatre Tourism

Akiko Sugawa-Shimada

Introduction

Theatre tourism is a type of tourism in which the destinations are theatres. It naturally means that the purpose of tourism for theatregoers – the 'tourist audience' (Bennet, 2005: 409) – is to see operas, plays, musicals and so on, performed at specific theatre buildings. The Paris Opera in France, the Kabuki Theater in Tokyo and the Takarazuka Theater in Kobe, for instance, are typical tourist destinations which offer high-quality operas and musicals all year around. Even site-specific theatre areas such as Broadway in New York and the West End in London are indicated in most tourist guidebooks. Therefore, theatre tourism seems to be one of the most popular forms of site-specific tourism. However, if the concept of contents tourism is applied to theatre tourism when the destinations are not site-and-area-specific, it sheds much light on the issues of 'contents' and fans' consumption.

Kontentsu tsūrizumu (contents tourism) is defined as 'travel behavior motivated fully or partially by narratives, characters, locations, and other creative elements of popular culture forms, including film, television dramas, manga, anime, novels, and computer games' (Seaton *et al.*, 2017: 3). Since in the Japanese cultural context contents tourism is alternatively called *anime seichi junrei* (anime-induced pilgrimage to sacred sites), specific sites and areas are often the focus of discussion. However, I argue that the potential of a *contents* tourism approach is that it focuses critical attention on the methods of consumption and use of *contents* by tourists/fans rather than the specificity of the places they head for as tourists. Tourism regarding 2.5-dimensional theatrical performances is a good example of this.

The 2.5-dimensional (in Japanese *ni-ten-go jigen*, hereafter '2.5-D') theatrical performances I discuss in this chapter are theatrical plays based on manga, anime and computer games. In Japan, 2.5-D theatrical performances have gained much attention from fans and within popular culture studies in recent years. The term 2.5-D refers to the space between two dimensions (the fictional space where our imaginations and fantasy work) and three dimensions (reality, where we physically exist). Owing to the growing success of cosplay and 2.5-D stage performances (theatrical adaptations of manga, anime, and games) since the early 2000s, 2.5-D has begun to be used as the umbrella term to express virtuality embodied by actual human bodies as well as human bodies that look unreal (I call this 'virtual corporality') in performances based on fictional images of manga, anime and video games.

In Japan the term *ni-ten-go jigen* (often abbreviated to simply 2.5, *ni-ten-go*) is closely associated with anime. In the 1970 and 1980s during the second anime boom,[1] 2.5 originally referred to animations' voice actors (*seiyū*). Regarding the functions of *seiyū*, Nozawa (2016) writes:

> The metaphor of *naka no hito* [person inside] frames *seiyū* as mediators or mediums of the character-driven convergence culture. It conjures up an imagination of the verge, where reality and fantasy meet but never merge. As subcultural participants often say, *seiyū* are seen as inhabiting an interstitial dimension between '3D' reality and '2D' fantasy – '2.5D'. *Seiyū* are mediums on the verge. (Nozawa, 2016: 170)

It is true that *seiyū* have an important function linking fantasy to reality with the help of the corporality of *seiyū*'s bodies and voices. Yet, in the digital era since the late 1990s, the ways in which fans consume and produce popular culture have drastically changed. One of the main characteristics in the consumption and production of anime, for instance, is the autonomy of anime characters within the *media mix* sphere (multiple media franchises). One of the best-known TV anime, NARUTO (2002–2007), for example, is based on Masashi Kishimoto's manga of the same title. It was adapted into 2.5-D theatrical performances, a Kabuki play, video games and arcade games. Numerous people cosplaying as NARUTO characters can be seen at any cosplay event worldwide. The images of its protagonists are used for merchandizing and other products, ranging from snack packages to posters for tourism promotion (such as in the character's namesake, Naruto city in Tokushima prefecture). The multimedia franchise can apply to any anime. Steinberg says about the Japanese media mix:

> [M]edia mix works as a system of objects and factors in media based on certain characters, narratives, and worlds (*sekai-kan*). [...] Media mix is the feedback system between consumers and producers, which has developed, thereby making consumers into producers. (Steinberg, 2015: 35)[2]

He also argues that the Japanese media mix seems to be a counterpart of North American transmedia storytelling. The existence of the characters is more powerful, but the continuum of narrative worlds is relatively flexible in the Japanese media mix (Steinberg, 2015: 25). Azuma also suggests that the media mix is a premise of database consumption in Japanese anime-related culture (Azuma, 2001: 63). These remarks typically account for one of the reasons for the emerging 2.5-D culture. Narrative worlds and characters are often considered as *kontentsu* (a loan word from the English 'contents'), namely 'information that has been produced and edited in some form and that brings enjoyment when it is consumed' (Okamoto, 2013: 40–41; see also Yamamura's Introduction to this volume). The flexibility of the narrative worlds in the media mix allows various expressions of characters and narratives to exist on multiple media platforms, which appeals to fans of anime, manga, games and light novels.

Regarding the tourists within contents tourism (or anime-induced tourism), Okamoto explains how fans enjoy consuming anime contents by migrating between three spaces: the fictional space of anime narratives and characters; the cyber space, where fans exchange information about stories and the stages of the stories; and the physical space, which comprises 'anime's sacred sites' for fans as well as the sites where local people spend their lives (Okamoto, 2015: 51). This model is applicable to 2.5-D culture. Fans as active consumers and producers concurrently access these spaces and take multiple levels of pleasure. In addition, reality and fantasy have been blurred through the development of visual technology such as virtual reality (VR) and augmented reality (AR), as well as users' communication through social media. These changes in visual technology and communication style affected our perception of 'the real'.

This chapter first explores the three major factors behind the emergence of 2.5-D culture in Japan: (a) our changing perceptions of 'reality'; (b) character-oriented consumption induced by the full normalization of the media mix (franchises in multiple media forms); and (c) fans' participation in the 2.5-D culture. It then argues how attending 2.5-D theatrical performances can be a form of contents tourism, and can be interpreted as a new type of site-non-specific theatre tourism through analysis of the ways in which Japanese and international fans have used two of the hit 2.5-D theatrical performances: *Musical Prince of Tennis* (2003) and *Token Ranbu* (2015).

Major Factors Within Emerging 2.5-D Culture

The development of visual technology and a change of communication style

The emerging 2.5-D culture is highly affected by the changing media environment and changes in our perceptions of reality. Virtual reality (VR) was often used to express a fictional world that resembles 'reality' when video games using VR were first introduced. Augmented reality

(AR) is a live, direct or indirect view of a physical, real-world environment whose elements are supplemented by computer-generated sensory inputs such as sound, video, graphics or GPS data. With the development of smart phones, tourist information or city guides, for instance, are often offered through AR-based software installed in mobile devices. With our smart phones, we can now easily experience AR images while travelling. Milgram and Kishino (1994: 1321) use the term 'Mixed Reality' to examine how 'both "virtual space" on the one hand and "reality" on the other [are] available within the same visual display environment'. De Sauza e Sylva suggests:

> Hybrid spaces are mobile spaces, created by the constant movement of users who carry portable devices continuously connected to the Internet and to other users. A hybrid space is conceptually different from what has been termed mixed reality, augmented reality, augmented virtuality, or virtual reality. (De Sauza e Sylva, 2006: 261)

As she notes, in the 2010s the dissemination of the internet, mobile devices like smartphones and social media like Twitter have affected our perception of the 'reality' and communication styles. Regarding VR experiences and a sense of embodiment, Bailenson (2018: n.p.) argues that we could be readily convinced that a virtual body belongs to us when we see ourselves in the virtual reality world. Young people growing up with this hybrid reality environment may have a strong perception of blurred reality and fantasy. For those who are seamlessly connected to cyber space 24 hours a day, physical and virtual contacts are not distinguishable.

Media mix

The Japanese term *media mikkusu* (media mix) was coined in the 1980s after Haruki Kadokawa, the then president of Kadokawa Books publishing company, established a successful business model using the media mix (Steinberg, 2015: 39). Kadokawa Books is the pioneering publishing company that introduced a media mix system in Japan. As a sales promotion campaign, they produced live action movies based on novels, such as the film *The Inugami Family* (1976) and *Sailor Suit and Machine Gun* (1981) (Ōtsuka, 2012, 2014). However, according to Ōtsuka, a system similar to the media mix can be traced back to the Edo period in the late 17th century and the early 18th century, when ancient war tales about famous historical figures were adapted into Kabuki theatre, and illustrations of the actors playing the protagonists were printed and sold in paper form (Ōtsuka, 2012: 45–49).

Thus, the media mix system has existed in Japan for a long time. Since TV anime, in particular, offers visual images of fictional characters animated with human voices, child TV viewers can feel intimacy with the anime characters as if they are alive. Furthermore, the ways in which TV anime are broadcast by private TV networks facilitates and naturalizes the seamless worlds of contents. In Japan, TV anime is mostly sponsored by companies which utilize merchandizing, so a 30 min slot of a weekly TV anime is divided into three parts: the A part (the first half of the episode), advertisements and the B part (the second half of the episode). Some advertisements use visual images of anime characters, which is not prohibited (unlike in the USA). As Azuma (2001: 63) indicates, the media mix is so disseminated in Japan that consumers pay less attention to the distinction between the original and copies.

In the media mix, anime characters' visual images are focused on more than the narratives they belong to. That is, the visual images of the characters can be seen not only in entertainment products but also in our daily commodities such as snack packages, clothing, stationary, and so on. As the change of our perception towards reality is influenced by the development of digital technology (the internet, social media, computer graphics, VR/AR etc.), the media and cultural environment promoted by the media mix can allow us to relate fictional characters to our daily lives. Although the media mix or 'convergence of multiple media platforms' (Jenkins, 2006: 14) environment is not exclusive to Japan, the Japanese media mix is highly disseminated and widely spread over a range from commercial works to fans' derivative works.

Fans' active commitment

The participation of fans is very important for constructing a 2.5-D space inside and outside the theatre. Fischer-Lichte (2014: 18) suggests that 'a performance is inseparable from the bodily co-presence of various groups of people who come together as actors and spectators'. She argues that any events such as wedding ceremonies, football games and street performances can be viewed as performances where there is active involvement of the audience as players, and interactive practices between actors and audiences exist.

Her argument is also applicable to 2.5-D theatrical performances and fans' cultural practices in terms of fans' deep commitment to the performances and their practices both in the physical and cyber spaces. In the cyber space, fans often check the Twitter feeds of the actors who are cast for a certain 2.5-D performance. Since most fans have read the manga, watched the anime and played the video games that the performance is based on, they have certain visual/vocal images prior to going to the theatre where the play is performed. Having these visual/vocal images, fans tend to enjoy how actors make efforts to copy the character whose role they play, and how the relationships between the actors differ from the ones between characters the actors play (Sugawa, 2018: 132).

Fans' communication through video clips streamed on YouTube and/or Niconico dōga (a Japanese video-sharing site) is also considered as part of their cultural practices. In the early period of *Musical Prince of Tennis*, some fans produced video clips ripped from DVDs with unique subtitles added to the clips, called *soramimi* (mishearing), and uploaded them to YouTube and Niconico dōga. Since most of the young cast members in *Musical Prince of Tennis* were inexperienced and have no history as professional musical actors, their immature actions and poor enunciation served as materials for fans to make fun of. Yet at the same time, the immaturity is viewed as a margin for growth, allowing fans to encourage cast members and take pleasure in seeing them mature into fully fledged actors (Sugawa, 2018: 136). Niconico dōga has a function by which users can leave and inscribe their comments in other users' movie clips, so fans are able to communicate with other fans through comments. In the early period of *Musical Prince of Tennis*, there was no nationwide tour, so manga and anime fans in rural areas became interested in this musical by accessing fans' *soramimi* clips. Such interactions facilitate the construction of a fandom in the cyber space, and eventually served to prompt fans to go to theatres in urban areas.

In the physical space, fans go to the theatre to appreciate performances. At the theatre, they purchase merchandizing products such as lightning sticks, which are used in revue show parts during the performance (see also Jang, Chapter 7). Some fans also exchange unwanted

Figure 8.1 Fans trade unwanted products and tickets outside the theatre. Author's photo

goods and tickets for the ones they need with other fans to expand their network Figure 8.1.

In all these ways, the development of visual technology, the dissemination of the media mix system, and fans' active participation in 2.5-D theatrical performances are interconnected to establish the current phenomenon of flourishing 2.5-D culture.

2.5-D Theatrical Performances and a New Type of Theatre Tourism

Emerging 2.5-D theatrical performances

2.5-D theatrical performances include musicals, straight plays and other forms of plays based on Japanese manga, anime and video games. The beginning of 2.5-D theatrical performances is probably *The Rose of Versailles*, based on Riyoko Ikeda's manga and performed by the Takarazuka Revue in 1974. The Takarazuka Revue is the Japanese all-female musical troupe founded in Hyogo Prefecture in 1914. *The Rose of Versailles* was based on the *shōjo* (girls') manga of the same title about the French Revolution and was an instant success.[3] Although Takarazuka actresses performed the manga characters in elegant costumes and with the hairstyles of 18th-century France, it was generally regarded as not 2.5-D but a Takarazuka production owing to the emphasis on the performance style of Takarazuka and the Takarazuka brand.

Since the Takarazuka Revue's *The Rose of Versailles*, several theatrical performances based on manga, anime and video games have been produced, although the number of the titles did not begin to rise suddenly until 2008, and the big leap came in 2012 with approximately 1000 titles (Pia, 2018). The precise reproduction of fictional characters and reenactment of the narrative worlds in manga, anime and video games were emphasized. For instance, in the serials of *Sakura Wars* (1997–2018) and *Hunter × Hunter* (2000–2004), the performers were the voice actors/actresses of the TV anime of the same titles (Yoshioka, 2018: 52). Yet, the 2.5-D stage scene drastically changed in 2003 when *Musical Prince of Tennis* started. The casts were made up of young, largely unknown actors. With the help of makeup, hairdressing and costumes, they visually resembled the manga/anime characters. The actors copied precisely the voices and behaviours of their characters in the anime version. These factors allowed audiences to put more focus on which characters were performed, not on who performed the character, since fans had less background knowledge about the actors.

According to a myth among 2.5-D fans, one fan of *Musical Prince of Tennis* in the early 2000s wrote a comment on social media saying, '[the cast] just looked like the characters as if they had popped up from 2-D [anime]. They are just 2.5-dimensional!' The complete reproductivity and reenactment of the original anime were highly appreciated by fans, and the performances came to be regarded as 2.5-D. This fan-led term has disseminated widely throughout the internet. In 2014, when the Japan 2.5-Dimensional Musical Association, which promotes any theatrical performances based on manga, anime and video games, took the term 2.5-D in its name, 2.5-D began to refer to not only representations of anime/manga characters performed by people's bodies, but also cultural products and performances based on fictional anime, manga and video games. As the number of titles of 2.5-D works increases, audiences have ever more opportunities to go to the theatre and be absorbed into the 2.5-D world.

Fans' theatre tourism

As I mentioned in the Introduction, theatre tourism generally requires specific theatre buildings and areas. However, there are no theatres dedicated only to 2.5-D theatrical performances except AiiA 2.5-D Theater in Shibuya, Tokyo. This theatre, however, closed in 2019 owing to the expiry of its contract. Usually in theatre tourism, a play runs at the same theatre for a long time – months, years or even decades in the case of the record-breaking run of *The Mousetrap* in London. However, 2.5-D theatrical performances are not stand-alone productions, but like anime and manga have multiple episodes within a series (called a 'season' in Japanese). Taking *Musical Prince of Tennis* as an example, the first 'season' ran from 2003 to 2010, with 16 plays ('episodes') in which each had a run of between a few days and a few months (*Musical Prince of Tennis*, n.d.). The entire season is not necessarily performed at the same theatre. The musical 'episodes' are interspersed with other events or concert performances. In a similar manner, the second season ran from 2011 to 2014, and the third season from 2015 to 2019. In the Kanto area, each play is usually performed at Tokyo Dome City Hall or Nippon Seinenkan Hall, and concerts are held in Pacifico Yokohama. The performances then go on a nationwide tour for about a week each in the Kansai, Kyushu, Tohoku and Chugoku areas. The show even goes on international tours to Asian countries/regions such as Taiwan and South Korea. Similarly, the performances and shows of the blockbuster serials of *Token Ranbu* (both musicals and straight plays) are given in several different theatres and event halls in Japan and overseas.

Consequently, 2.5-D performances closely fit the notion of contents, not only because there are multiple stage performances drawing on the same narrative world, but also because the narrative world itself is disseminated across various media platforms including the stage, manga, anime and concerts. Fans of a production like *Musical Prince of Tennis* have had the opportunity to not only see the same episode multiple times, but also to see different episodes on an almost monthly basis. Videos of particular episodes are sold on DVD too. Given a loyal fan base and considerable opportunities for repeat viewings, *Musical Prince of Tennis* had amassed a combined audience of over 1.7 million theatre-goers in its first 10 years (Animeanime, 2014).

I argue that 2.5-D theatre tourism does not require specific locations and areas. It can be carried out in various spots where the contents are present (see also Hood, Chapter 11). The various 'episodes' of *Musical Prince of Tennis* and *Token Ranbu* are not all performed at a single specific theatre, so the theatre itself does not become a fixed 'sacred site' for fans of 2.5-D theatrical performances. Therefore, the space where a performance takes place, and where fans can gather temporarily, becomes a cluster of 'sacred spots' for fans. Here the term 'spot' is more appropriate than the better known phrase 'sacred site' (*seichi*) because there are multiple spots where fans gather, such as near the main entrance, or before the life-size cardboard cutouts of the casts where fans take photos.

Such 2.5-D theatrical performances have also induced inbound tourism, for example *Live Spectacle NARUTO*. *NARUTO* is one of the most popular set of contents within Japanese pop culture. There are numerous fans of *NARUTO* in its manga, anime and video game versions all over the world. The first episode of the serialized plays of *Live Spectacle NARUTO* was performed in Japan in 2015. Although *Musical Prince of Tennis* went on tour in Taiwan and South Korea prior to *NARUTO*, *Live Spectacle NARUTO* was the first acknowledged 2.5-D theatrical performance that went on a full-scale international tour.

Prior to the international tour, the final performance of the first play in the series of *Live Spectacle NARUTO* in Japan was streamed live in movie theatres in Japan, Hong Kong and Taiwan, a practice called 'live viewing'. This prompted international fans of *NARUTO* to go to theatres and see the real performance. In China, live viewing is not available under the country's restrictions on the media and internet. However, *NARUTO*'s fans access information through the internet. Most young fans are familiar with the Virtual Private Network system, and they use an application to exclude the access limiter. Fans were able to access clips from DVDs uploaded by anonymous fans on streaming sites and/or information about *NARUTO* on Facebook and Twitter, which are supposed to be inaccessible in China. Therefore, the international tour of *Live Spectacle NARUTO* in Macau, Singapore, and Malaysia in 2015, and China and Malaysia in 2016 ended up successfully (Animation Business Journal, 2017). In China, the tour covered Shanghai, Hangzhou, Beijing, Changsha, Guangzhou and Shenzhen. Some enthusiastic fans followed the troupe to visit every city in China. Additionally, enthusiastic international fans have come to Japan to see other 2.5-D theatrical performances that are not performed in the countries where they live. Some of them have even settled down in Japan specifically to have more chances to go to see 2.5-D performances. Evidence of this, what one might call 'contents migration', is given by some of my Chinese informants as discussed below.

A similar situation can be observed regarding the serials of *Musical Token Ranbu*. It is based on a hit online game, Token Ranbu Online (2015), in which Japanese swords that historical figures possessed are personified as a hunk. In the media mix, it was adapted into two TV anime, one live action film, two serialized theatrical performances (musical and straight play) and live concerts. Sales of merchandise have been brisk. Because it is a character-growing game and action game, there is no specific storyline. Fans enjoy creating diverse stories using the settings and characterizations of the game's characters.

The musical version and its related shows were staged in several cities in China, and most recently in France. Some enthusiastic fans living in Japan followed the international tour, and some international fans of *Token Ranbu* came over to Japan to see the performances only available in Japan, contributing to inbound tourism (NHK, 2018).

Fans as active players

One characteristic of 2.5-D theatre tourism is the fans' activities inside and outside the theatres. Inside the theatre, fans purchase novelty goods such as badges, photographs of the casts and light sticks to prepare for the performance they will see. Some performances such as *Musical Prince of Tennis* and *Musical Token Ranbu* have a concert and a talk show after the play for audiences to participate in. The audience members swing light sticks and sing songs together during the concert and the show. Before and after the performance, and during the break, members of the audience often type their comments and messages on Twitter.

Outside the theatre, before and after the performance fans exchange unwanted merchandise for items they want. There is a way of selling products to fans called 'random goods'. Merchandise such as photos and badges are sold in sealed paper bags. Since fans never know what items they will get, fans purchase the random goods as if they are playing the lottery. They tend to continue buying merchandise until they obtain the particular items they want. As a result, they have many extra unwanted items that they trade. They also resell extra tickets to fellow fans after making appointments to meet up through social media like Twitter (see Figure 8.2).

Therefore, the areas inside and outside of the theatre become 'sacred spots' for fans to communicate and interact with other fans and become more engaged with the theatrical performance. Although the places are not fixed, wherever the contents are, the space where 2.5-D theatre fans gather becomes a destination of 2.5-D theatre tourism.

Figure 8.2 Fans waiting outside the theatre in a 'sacred space' of theatre tourism. Author's photo

Qualitative Research on Fans' 2.5-D Theatre Tourism

A focus group discussion with Chinese fans

On 30 April 2018 I organized a focus group discussion with six Chinese 2.5-D theatre fans who live in Tokyo. I met two of them at the theatre, and they invited their friends to this discussion. All of the informants have seen Japanese 2.5-D theatrical performances, such as *NARUTO*, *Musical Black Butler*, and/or *Musical Token Ranbu* in China. Some of them had travelled repeatedly to Japan from China specifically to see performances available only in Japan. Ultimately they have all settled in Tokyo, where they are either students or work full-time (Table 8.1).

Informant A loved Japanese anime such as *Cardcapters* and *Neon Genesis Evangelion*. She saw *Musical Sailor Moon* on DVD and went to see *Musical Black Butler* at the theatre in Beijing in 2015. She immediately became a fan of 2.5-D theatrical performances, and then decided to go to Japan to study at graduate school. Similarly, Informant B loved the anime *Sailor Moon* as a child. She watched *Musical Prince of Tennis* on DVD. She did not like it at first, but once she saw *Musical Black Butler* at the theatre in Beijing, she was overwhelmed. At that time, she fell in love with a Japanese actor, Yukito Nishii, who was in the 2010 Japanese movie *Kokuhaku* (Confession). Her enthusiasm for good-looking Japanese actors brought her to Japan. Other informants have similar experiences

Table 8.1 Background of the six informants in the focus group discussion

Informant	Age	Occupation in Japan	Duration of Stay
A	29	Student at a vocational school (studying design and illustration) in Tokyo	1.5 years
B	25	Graduate student at a private university in Tokyo	1 year
C	29	Graduated from the graduate school of a private university in Tokyo. Working for a game company	7 years
D	27	Graduate student at a private university in Tokyo	6 years
E	—	Graduate student at a private university in Tokyo	3 years
F	26	Graduate student at a university in Singapore. In Japan temporarily for research	0.6 years

viewing anime in their childhood. Then around 2010, they experienced 2.5-D theatre in China. Some of them often came to Japan as temporary visitors to see performances not available in China. Although all of them came to study in Japan, their primary purpose was to attend 2.5-D performances and events. This exemplifies the potential of contents tourism to be a step towards 'contents migration', or migration induced by popular culture content.

Another interesting point they share in common is the way in which they consume and use Japanese popular culture besides anime and manga. Informant C shifted her interest from Japanese anime to male *aidoru* singers of the famous Johnny's Co. Ltd artist management agency. In the Japanese cultural context, *aidoru* (or pop idols) signifies young, inexperienced, but promising singers who are cute and/or cool. Since their appearance and ephemerality are usually the focus of attention, their popularity is mostly among young people. When she found similar pleasurable aspects of idol consumption within 2.5-D performances, she instantly became a great fan of 2.5-D theatrical performances. By the time she finished her undergraduate course at a Chinese university, she had decided to come to Japan as a graduate student, expecting to have more chances to see 2.5-D performances. She successfully made friends with other fans through trading merchandise outside the theatre. Taking advantage of her stay in Japan, she often purchases merchandise at 2.5-D theatrical performances to send to Chinese friends who she has met through social media in China. In other cases similar to hers, merchandise only available for purchase at the theatres strongly motivates fans to be 2.5-D theatre tourists.

Similar experiences were shared by Informant E. Informant E first became interested in Japanese pop music in high school, and then expanded her interest to Japanese films and novels in her university days. Since she usually consumed Japanese pop cultural products, she also watched Japanese anime on the internet. When she fell in love with the voice of a character in the anime *Terror in Tokyo* (2014), played by the emerging voice actor Soma Saito, she began to attend 2.5-D theatrical performances because of their connection with anime. Her first experience of a 2.5-D play was *Musical Token Ranbu* in 2015. She began to play the game too. Informant F had a similar cultural experience. She grew up with Japanese pop culture and became interested in 2.5-D theatre because she was also fond of any kind of theatrical performance. She preferred Japanese 2.5-D performances because of the use of projection mapping and other high-tech stage sets.

Cosplay also connects these fans to 2.5-D theatre tourism. Informant D was an active cosplayer and belonged to the theatrical club in high school. Informants A, B, C and E used to cosplay in China, too. Informants A and B commented:

> I loved cosplaying. In 2.5-D performances, I can see two-dimensional characters embodied as three-dimensional persons. I found some similarities to cosplay. (Informant A)

> I first got to know about 2.5-D via an article on *Tenimyu* [*Musical Prince of Tennis*] in a Chinese anime magazine, *Man'yu*. As I also enjoyed cosplaying at that time, I immediately accessed to the Internet to gather more information. I finally got a ticket to see the real *Musical Black Butler* at the theatre in China. It was just awesome! I liked their costumes. (Informant B)

As their comments indicate, cosplay culture has taken root among young Chinese women, and they seemed to regard 2.5-D theatrical performances as a sophisticated version of cosplay. Since the informants felt that they were too old to be cosplaying anymore (a theme also discussed by Rastati, Chapter 9), they simply take pleasure in seeing young actors dressed like anime characters as a substitute behaviour. This prompts them to go and see 'real' (meaning 'professional') 2.5-D theatre.

Individual interviews with weekend 2.5-D theatre tourists

I also interviewed two international fans who came to Japan to see 2.5-D theatrical performances on weekends in April and May 2018. I would call them 'international weekend 2.5-D theatre tourists'. One is a Chinese woman in her twenties who is a full-time businesswoman in China and often comes to Japan to see 2.5-D plays at the weekend. Usually there are two performances (matinee and soiree) of a single title on weekends. Therefore, she attempts to see four performances of the single title (two on Saturday and two on Sunday). She rarely has free time for sightseeing, but on occasions when she is unsuccessful getting a theatre ticket, she does do some sightseeing, such as visiting Osaka Castle. She is a great fan of Taishi Sugie, one of the top 2.5-D theatre actors, and has made friends with many Japanese fans of Taishi through his fan club on social media. She keeps her weekend tours to Japan a secret from her work colleagues because 'they hardly understand [her] taste'.

An Australian female college student in her twenties has a similar situation. When I interviewed her, she told me that she flew to Japan from Sydney to see a play from the serial *Hyper Projection Play, Haikyu!!* (2015) for the first time because she got a three-day holiday for the weekend before her mid-term exams. She did not tell her parents and friends about her trip to Japan. She is afraid of others thinking her 'strange' because manga, anime and games are not for adults but children in her culture. She first fell in love with the anime *Haikyu!!*, and then started to read the manga on which the anime was based. However, when she heard of the theatre adaptation of *Haikyu!!* and saw a movie clip on YouTube, she said she just had to see a 'real' performance in Japan. Identifying herself a 'beginner' fan, she belongs to no specific fandom yet. However, she was hoping to return to see the new episode of *Hyper Projection Play Haikyu!!*

In both cases, these women share a strong sense of isolation among friends at work (for the Chinese woman) and university (for the Australian girl). They have no friends nearby with whom to share their hobby. In fact, this is the case for most 2.5-D fans: they hesitate to speak out about their hobby and seek people outside their day-to-day communities who share and understand their tastes and preferences. The theatres serve as sites for fans to find someone with similar tastes. This is one of the key factors motivating 2.5-D theatre tourism.

Conclusions

The most attractive characteristic of Japanese 2.5-D theatrical performance is that fictional characters of anime, manga and video games come alive via the human body with a high level of reproductivity and reenactment. The dissemination of the media mix environment, the change of our perception of 'reality' owing to the development of digital technology and fans' active commitment to cultural practices have made 2.5-D culture possible in Japan. In this media milieu, tourism induced by Japanese popular culture has often been called 'pilgrimage to sacred sites' (Seaton *et al.*, 2017) with more emphasis put on 'sites/places'. However, when contents, consisting of the characters and the narrative world, are more the focus of attention, any 'spots' in the physical space where the contents and fans come together may become a destination for tourism within the framework of contents tourism; 2.5-D theatre tourism exemplifies this.

Fan communities attract many other fans who share and understand similar tastes, which serves to facilitate 2.5-D theatre tourism. Besides enjoying the fictional spaces in performances at theatres, fans are connected by the same or similar preferences both through social media, like Twitter, in the cyber space and through interaction inside and outside the theatre. They often communicate with one another by trading character merchandise, such as photos and badges, or by reselling extra tickets.

They post on Twitter and become acquainted inside and outside of the theatre. By so doing, they expand their networks. I would call this 2.5-D fandom a series of 'communities of preferences' with a stress on the word 'preference'. They are what Tomita (2017: 156) calls 'intimate strangers' because they generally hide their identities and backgrounds. However, this superficially shallow distance is comfortable for fans.

Thus, 2.5-D theatre tourism is a new type of theatre tourism and can be considered as an example of contents tourism. There are no fixed locations and places that can be regarded as the 'sacred sites' of 2.5-D theatrical performances. However, any spots inside and outside the theatre can be 'sacred' for them if they can meet their favourite fictional characters, favourite actors and other fans. It is true that there are similar fandoms for Broadway musicals, Takarazuka and Kabuki that can be observed as forms of theatre tourism. However, when the focus is on the contents, not only the appreciation of the performances but also the interactions with other fans become the essential characteristics of 2.5-D theatre tourism.

Notes

(1) The first anime boom in Japan was induced by *Tetsuwan Atom* (*Astroboy*, 1963–1966). The second anime boom was induced by the movie *Uchū Senkan Yamato* (*Space Cruiser Yamato*) in 1977 (Tsugata, 2017: 84).
(2) Although this is a from a translation of *Media Mix: Franchising Toys and Characters in Japan* (Minneapolis, MN: University of Minnesota Press, 2012), it is the revised and updated Japanese version. The citation here is not in the original book, and thus, it is my translation.
(3) The musical was performed until 1976, marking a record of over 1.4 million visitors (Takarazuka Revue Official Website, n.d.). *Rose of the Versailles* (1972) was adapted into a TV anime (1979–1980) and a live action movie (1979). Its media mix products are various.

References

Animation Business Journal (2017) Sutēji ban 'NARUTO' kaigai 6-kakoku chiiki de raibu byūingu, kaigai ni hirogaru 2.5-jigen. See http://animationbusiness.info/archives/3249 (accessed December 2018).
Animeanime (2014) Tenimyu 10-shūnen, shuku! Ruikei kankyaku dōinsū 170-man-nin toppa, kōen & atorakushon @ Tōkyō dōmu shiti. See https://animeanime.jp/article/2014/02/08/17381.html (accessed January 2019).
Azuma, H. (2001) *Dōbutsuka suru Posutomodan: Otaku kara mita Nihon Shakai*. Tokyo: Kōdansha.
Bailenson, J. (2018) *VR wa Nō wo Dō Kaeru ka? Kasō Genjitsu no Shinrigaku* (trans. Y. Kurata) (Kindle). Tokyo: Bungei Shunju.
Bennet, S. (2005) Theatre/tourism. *Theater Journal* 57 (3), 407–428.
De Sauza e Silva, A. (2006) From cyber to hybrid: Mobile technology as interfaces of hybrid reality. *Space and Culture* 9, 261–278.
Fischer-Lichte, E. (2014) *The Routledge Introduction to Theatre and Performance Studies*. London: Routledge.
Jenkins, H. (2006) *Convergence Culture: Where Old and New Media Collide*. New York: New York University Press.

Milgram, P. and Kishino, F. (1994) A taxonomy of mixed reality visual displays. *IEICE Transactions, Information and Systems* E77-D, 1321–1329.
Musical Prince of Tennis (n.d.) Kore made no kōen. See https://www.tennimu.com/archive/ (accessed January 2019).
NHK (2018) Shibuya nōto presents myujikaru Token Ranbu: 2.5-jigen kara sekai e (broadcast 27 October).
Nozawa, S. (2016) Ensoulment and effacement in Japanese voice acting. In P.W. Galbraith and J.G. Karlin (eds) *Media Convergence in Japan* (Kindle) (pp. 169–199). Kinema Club.
Okamoto, T. (2013) *n-th Creation Tourism: Anime Seichi Junrei/Kontentsu Tsūrizumu/Kankō Shakaigaku no Kanōsei*. Ebetsu: Hokkaido Bokengeijutsu Shuppan.
Okamoto, T. (2015) Kontentsu tsūrizumu no kūkan. In T. Okamoto (ed.) *Kontentsu Tsūrizumu Kenkyū: Jōhō Shakai no Kankō Kōdō to Chiiki Shinkō* (pp. 50–51). Tokyo: Fukumura Shuppan,
Ōtsuka, E. (2012) *Monogatari Shōhiron Kai*. Tokyo: Ascii.
Ōtsuka, E. (2014) *Media Mikkusu ka suru Nihon*. Tokyo: East Shinsho.
Pia (2018) Zennen hi 21% zō. Kyū seichō no 2.5-jigen myūjikaru shijō/pia sōken ga chōsa kekka wo kōhyō. *Pia*, 1 August, n.p. See https://corporate.pia.jp/news/detail_live_enta20180801_25.html (accessed December 2018).
Seaton, P., Yamamura, T., Sugawa-Shimada, A. and Jang, K. (2017) *Contents Tourism in Japan: Pilgrimages to 'Sacred Sites' of Popular Culture*. Amherst, NY: Cambria Press.
Steinberg, M. (2015) *Naze Nihon wa 'Media Mikkusu suru Kuni' nanoka* (trans. Y. Nakagawa). Tokyo: Kadokawa.
Sugawa, A. (2018) Ōdiensu, fan ron (fandamu): 2.5-jigen ka suru fan no bunka jissen. In M. Koyama and A. Sugawa-Shimada (eds) *Anime Kenkyū Nyūmon, Ōyō hen: Anime wo Kiwameru 11 no Kotsu* (pp. 118–142). Tokyo: Gendaishokan.
Takarazuka Revue Official Website (n.d.) Takarazuka kageki no ayumi (1962–1981 nen). See https://kageki.hankyu.co.jp/fun/history1962.html (accessed January 2019).
Tomita, H. (2017) *Intimate Stranger: Tokumeisei to Kōkyōsei wo meguru Bunka Shakaiteki Kenkyū*. Osaka: Kansei Daigaku Shuppan.
Tsugata, N. (2017) *Animēshon Gaku Nyūmon* (new edn). Tokyo: Heibonsha.
Yoshioka, S. (2018) The essence of 2.5-dimensional musicals? *Sakura Wars* and theater adaptations of anime. *Arts* 7 (4). See www.mdpi.com/2076-0752/7/4/52/htm (accessed December 2018).

9 Indonesian Cosplay Tourism

Ranny Rastati

The Emergence of the Indonesian Cosplay Phenomenon

The first Japanese anime that aired in Indonesia was *Wanpaku Omukashi Kum Kum*, which was broadcast on Televisi Republik Indonesia in the late 1970s (Rastati, 2012: 4). However, it was not well received. However, in 1986, the Japanese drama *Oshin* was highly rated by Indonesian viewers. Indonesians gained an image of Japan as a high-tech country that still embraced traditional values. The number of Japanese television programmes broadcast in Indonesia has increased since the establishment of the private television stations Rajawali Citra Televisi Indonesia in 1987 and Indosiar in 1991. Anime such as *Doraemon* and *Sailor Moon* were aired in the early 1990s and were popular among school-age children. Over time, the generation of the 1990s started bringing their favourite characters into the real world via cosplay.

This chapter discusses the potential of cosplay as contents tourism in Indonesia and by Indonesians, using the understanding of cosplay in Takayoshi Yamamura's introduction to this volume. It is based on in-depth interviews with nine Indonesian cosplayers, observation at cosplay events during 2017 and 2018 in Jakarta, Depok, Bandung and Padang, and secondary resources on official websites, news sites and fans' online sites (topics relating to cosplay culture and contents tourism in Indonesia are not widely discussed in academic sources). When cosplay started in Indonesia in the early 2000s, it was oriented towards Japanese cosplay owing to the popularity of manga and anime. The majority of Indonesian cosplayers used Japanese stage names like Kisame and Matcha Mei. It was rare to find cosplayers who used non-Japanese names, let alone their real names. Cosplayers use Japanese stage names to break away from their original identities (Rastati, 2012: 48–49). They want to look unique, stand out and be different from others. Japanese stage names also help the cosplayers to create a distance between themselves and the characters that they play. However, after 2010, cosplayers in Indonesia began to use combinations of their real names and non-Japanese stage names, such as Rian CYD and Rangga Kanchiel. The desire to express their identity as a cosplayer without hiding their real identity was one of the reasons.

In many regions, Japanese and cosplay festivals in Indonesia have been pioneered by Japanese studies student associations and supported by the Japan Foundation or Embassy of Japan in Indonesia. They include Gelar Jepang Universitas Indonesia or GJUI (University of Indonesia, Depok), UIN Matsuri (Sunan Gunung Djati State Islamic University, Bandung) and Japanvaganza (University of Hasanuddin, Makassar). Indonesian cosplayers mostly live in metropolitan cities such as Jakarta, Bandung and Makassar, and come from middle class urban youth in their twenties. Even though there are Indonesian cosplayers under the age of 20 and over 30, the numbers are not significant. Cosplayers, therefore, have similar backgrounds in terms of their level of education, economic situation, cultural tastes and consumption patterns. Some cosplayers state that they will stop cosplaying when they get married or reach the age of 30, at which point they feel that society expects them to act like adults and to take responsibility.

> If I get married, I want to stop cosplaying because I want to focus on my family. Cosplay is indeed a hobby that requires a lot of time and money. There is a lot of preparation ... I am afraid that I will neglect my husband and children. I have been cosplaying for years, so I am satisfied enough. My youth is well spent ... so I think it is time to be more serious in life. (Respondent W, 2018)

> If I marry, I am willing to continue to cosplay. This is not just a hobby but also how I make a living. I sell cosplay costumes and accessories, too. My girlfriend and family fully support me because I am able to show them that I am financially independent. Even though the money is not that much, I can make money from cosplay. But yes ... when I turn 30 I want to be more settled as a man. I have to support my family. As you know, society still considers cosplay as a child's hobby. (Respondent K, 2018)

Before 2010, cosplayers could be classified into two categories, junior and senior, based on their level of experience (Rastati, 2012: 43). Junior cosplayers were newcomers who cosplayed less than twice a year. They bought costumes because they lacked costume-making skills. Senior cosplayers, in contrast, had been cosplaying for years, attended more than two cosplay events a year and could make costumes. Some of them were cosplay competition winners and popular within the cosplay community. Furthermore, they were invited to be cosplay judges and guest stars in regional cosplay events.

Since junior and senior cosplayers were school or university students, they did not have a fixed income. Consequently, they could not afford to attend cosplay events abroad. Most of them used pocket money from their parents to make/buy costumes. They could not do part-time jobs, which were not widely available owing to the high number of people in the workforce. To overcome this, they sold old costumes to make/buy new ones. Those with the ability to make costumes could start a costume-making service to make money. Risa, a cosplayer from Bandung, remarked that costume-making for others is common because not all cosplayers have the skill to make their own (Rastati, 2017: 219–220). Another Bandung cosplayer, Makoto, added that he was able to get extra money from making costumes and some of his old costumes are still bought and sold within the cosplayer community.

> As time passed, I gained weight [laughs] and my old costume became so tight. Then I decided to sell it online. I sell a 'preloved' [second-hand or used] costume for around IDR 200.000 [around $15] to IDR 500.000 [around $38]. It's very cheap. Compared with the cost of making a costume, selling preloved items is not profitable, it's like a side job. But for me, being able to wear the costume several times and still managing to sell it makes me feel lucky enough and happy. (Makoto, 2017)

According to the Indonesian Internet Service Provider Association, the number of internet users in Indonesia increased from 30 million to 42 million people in 2009 and 2010 (APJII, 2010). However, the characteristics of cosplay changed after 2010, when social media became a major influence. That year, a new trend for Android-based smartphones in Indonesia replaced the previous trend for Blackberry (Nugraha, 2011). Indonesians were starting to access social media through applications on smartphones. Previously they had accessed it via computer. This led to a Facebook boom in 2010, especially among young people (Isadora *et al.*, 2012: 2–3). Android smartphone users, including cosplayers, can easily upload photos from smartphones to social media without the need to transfer photos from their cameras to their computers. Through social media, cosplayers are able to share photos and access information about cosplay events in other countries. Online interactions turned into offline friendships when they were able to meet at cosplay events. This led to a rise in cosplay collaboration between cosplayers from different regions.

The popularity of Instagram, meanwhile, has generated new advertising patterns. Business people no longer only use celebrities from the mainstream media and entertainment worlds to promote their products. Instagram celebrities or influencers are also asked to promote commercial products through endorsement and as brand ambassadors. Thanks to social media and sponsorship from the organizers, some senior cosplayers are able to earn an income and have been transformed into celebrity cosplayers. The presence of cosplay has also grown in the creative industries and screen culture. Skilled Indonesian cosplayers such as Franzeska Edelyn and Punipun were assigned as the official cosplayers of re:ON Comics, an Indonesian comic publisher. They also became brand ambassadors endorsing gadgets, notebooks and games. Cosplay, which was originally just a hobby, slowly turned into a profession and an industry offering many benefits and financial rewards to its members.

One of the first generation of Indonesian cosplayers is Pinky Lu Xun. Pinky started cosplaying in 1998 and her dedication led her to be dubbed the 'Queen of Indonesian cosplayers'. In 2003, she was desperate because she could not find another Indonesian cosplayer. However, in 2004 Pinky finally met another cosplayer named Orochi at the event Animonster Sound Aishiteru. She cancelled her retirement as a cosplayer and formed an all-female cosplay group named Endiru 'Endless Illution' together with Orochi and three other cosplayers named Shinji, Silver Yuna and Tsadhe. As cosplayers, Pinky and Orochi have received high recognition at both national and international levels. Cosplay communities recognize them as true cosplayers because of the quality of their costumes, makeup and performances on stage. They are often appointed as cosplay judges and appear as special guests at prestigious cosplay events. They also attend meet-and-greet events with local and international fans. Both of them are not only known as the leaders of the first generation of Indonesian celebrity cosplayers, but also as leaders of the first generation of Indonesian international cosplayers.

Social Attitudes Towards Cosplay

Although cosplay communities have increased in popularity, size and prominence, particularly in big cities, some sections of society still frown upon cosplayers. People who support cosplay consider it to be a creative activity. However, others consider cosplay to be strange, childish and a waste of money. In such circumstances, many cosplayers decide to hide their identities as cosplayers from parents, family and friends.

> My family and classmates don't know that I am a cosplayer. I am afraid that my dad will not allow me to do it if he finds out. Once, I asked him about cosplay when we saw a cosplay event on the news. He said it looked ridiculous. So, I don't have the guts to tell him more. When attending cosplay events, I wear my regular clothes first. Then, I change into my cosplay costume in my friend's house. My friend is a cosplayer too ... But she receives the full support of her family. (Respondent S, 2018)

Cosplay in Indonesia has several interesting characteristics. For example, crossdress (a male cosplayer as female character and vice versa) began to appear in Jakarta in 2004 and increased significantly in 2006–2009. The development of crossdress in the Indonesian cosplay community was triggered by the emergence of a female impersonator cabaret in the 2000s (Udasmoro, 2017: 320–321). A comedian called Tata Dado founded an all-male cabaret dance group, Silver Boys, who dressed up as females. This act split public opinion because Indonesia is predominantly Muslim and strictly against crossdressing. There is also a negative stigma attached to being a male crossdresser. With homophobic undertones, it is considered to be 'gay' or an 'unmanly pursuit'.

In 2010, religious (Islamic) awareness increased among young people. Religious preaching, which had been done mainly in houses of worship, began to be broadcast on television and YouTube. Young preachers were seen as both pious and modern/cool, and came to be idolized. This was a milestone in the modernization of Islamic values with Islamic teaching made through young preachers on social media (Jati, 2015: 155–159). As a result, in 2010 the number of crossdressers decreased, especially male crossdressers. According to Respondent R (2018), a male cosplayer in his late twenties, being a crossdresser now requires a strong mentality. For him, crossdress is not strange as long as it is part of acting and cosplay. And Rangga Kanchiel, a well-known cosplayer from Bandung, has said:

> Crossdress, in my opinion, is something creative because one can play the other side of oneself. If a male crossdresser is praised for being beautiful, he will be proud and satisfied and enjoy the attention given. For me it is an art. But, Indonesia has a Muslim majority, so some people are not comfortable to see men crossdressing. It is against the religion of Islam. God curses someone if he/she wears clothes that resemble the opposite sex. So, that is why only a few cosplayers are confident enough to crossdress. (Rangga Kanchiel, 2017)

After the decline in interest in crossdress, hijab cosplay emerged as a new phenomenon in 2012–2014 (Rastati, 2015: 380). Hijab cosplayers who took off their hijab when cosplaying are now able to use certain techniques to imitate the hairstyles of characters without releasing the hijab (Figure 9.1). Stylish hijab in Indonesia have spread rapidly since 2010 following the establishment of Hijabers Community, which was formed by young fashion bloggers and designers such as Dian Pelangi and Jenahara (Beta, 2014: 379–381). The popularity of the hijab has also reached fan communities like Japanese pop culture enthusiasts. Hijab manga and hijab cosplay are just two of the new phenomena that have emerged as a result of the widespread wearing of the hijab in public. The hijab was transformed from something seen as traditional into something deemed stylish and trendy. Hijab cosplay has grown in popularity, especially in Muslim-majority countries such as Indonesia and Malaysia.

However, hijab cosplay generated controversy. Hijab cosplayers received pressure from within the fan communities. Those who support hijab cosplay argue that cosplay is an activity for everyone regardless of race, age and faith. In contrast, others have stated that cosplay is an activity during which cosplayers should let go of their personal beliefs because cosplay is a time to show one's skill in bringing to life the appearance and behaviour of other characters. Furthermore, characters in manga and anime are rarely shown in religious settings. By adding religious symbols into the performance, hijab cosplay seems to drown the character in

Figure 9.1 Hijab cosplayers spotted at the Bandung Lautan Costume event on 7 May 2017. Author's photo

religion (Rastati, 2017: 216–217). As one hijab cosplayer supporter, Rio 'Kaicho', founder of Islamic Otaku Community, has stated:

> Hijab cosplay is in the gray zone. From the Islamic zone, we are rejected because enjoying our hobby of anime and manga is weird. Japanese pop culture lovers and the cosplay community also refuse to consider hijab cosplay as part of cosplay. We are *otaku* [people obsessed with Japanese pop culture] who still maintain Islamic values such as praying five times a day, covering awrah [body parts that cannot be seen by the opposite sex] and maintaining good attitudes when cosplaying. (Rastati, 2015: 382)

Another interesting characteristic of cosplay in Indonesia is nationalist awareness of Indonesian culture. Indocosu (an abbreviation of 'Indonesian and cosplay') is cosplay based on Indonesian characters such as Si Unyil and Wiro Sableng. This phenomenon began to appear in 2009–2010 (Rastati, 2015: 380) and developed significantly in 2013 along with the establishment of local publishers such as re:ON Comics, Koloni and Ngomik. Since the 1930s, comics in Indonesia have been mostly adapted and translated from publications in the USA and Japan (Lubis, 2009: 60–62). Japanese comics accounted for 90% of comic sales in Indonesia in the 1990s (Imanda, 2002: 49). However, the emergence of local publishers who publish Indonesian-themed comics based on national cultural values suggests that Indonesian comics are also gaining popularity.

Indonesian cosplayers, therefore, also have various choices for cosplay not only from foreign characters but also from those within the country.

Recently, there has also been an expansion in the definition of cosplay in Indonesia beyond dressing as characters from Japanese popular culture. Characters from the USA and China, such as Spiderman and Sun Wukong the Monkey King, are performed by Indonesian cosplayers. Moreover, cosplayers are starting to cosplay as non-pop-culture characters, such as security officers and mayors. There are also those who cosplay as actual people (cosplay impersonators) like politicians and comedians. This type of cosplay used to be very unusual in Indonesia. However, it seems that cosplay in Indonesia has been increasingly legitimized as an enjoyable activity which is flexible and able to move from one format to another without losing its essence as an identity game. Cosplay does not function as a simple form of escapism that allows cosplayers to escape their mundane lives, but is instead an important activity for the creation of identity among the fans of manga and cosplay (Bonnichsen, 2011: 2). The most common motivations for cosplayers include entertainment, escape from everyday life, belongingness, eustress and aesthetic beauty (Reysen *et al.*, 2018: 29). Anyone can participate in cosplay without being limited by strict rules that can damage the pleasure itself.

Cosplay as Contents Tourism

Contents tourism is a new concept in the Indonesian academic world. Contents tourism, as opposed to film-induced tourism and other such terms, is the term adopted by the Japanese government in its official promotional strategies (Seaton & Yamamura, 2015: 2). Popular culture products such as novels, movies and anime trigger fans to travel to places that appear in the popular works. Fans tend to be willing to spend significant amounts to consume various things related to the contents they like (Pratama, 2018: 115–116). This includes travel. The potential of cosplay tourism has been promoted since 2016 by Ridwan Kamil (Mayor of Bandung 2013–2018) (Perdana, 2016). Bandung is known as one of Indonesia's most creative cities and has held many cosplay events, such as Hanami Festival Cosplay and Bandung Lautan Costume. Street cosplayers dressed as Hello Kitty, Captain America and Indonesian ghosts such as Kuntilanak can also be found in Asia Afrika Street and Cikapundung Terrace Park, the tourism icons of Bandung (Figure 9.2). Cosplay has become a tourist attraction in Bandung since information was posted on the government website. It has a unique and high value that makes tourists willing to visit (Nuansya, 2017: 10). Ridwan Kamil has stated:

> Cosplay seems to be a unique tourism attraction in the city of Bandung. I support it and I do not forbid it. It will be an effort for Bandung City Government to manage tourism. The more creative and similar the cosplay costume, the happier visitors are. (Perdana, 2016)

Figure 9.2 Cosplayer in Asia Afrika Street, Bandung (April 2018). Author's photo

In addition to Bandung, Jawa Timur (Jatim) Park has also attracted cosplayers. Established in 2001 and located in Batu City, Jatim Park has an attraction called The Legend Park. The Legend Park has replicas of international landmarks, structures from movies such as Harry Potter and a Japan spot. According to the Operational Manager of The Legend Park, Tossy Kusdianto, the Japan spot is a favourite area because of its traditional Japanese houses and replica cherry blossoms (Richa, 2018). This same newspaper report also discussed the presence of 12 cosplayers on a certain day, dressed as characters from the anime *Naruto* and entertaining visitors, who could take pictures with the cosplayers and rent cosplay costumes.

The popularity of Japanese pop culture products made Japan the country that Indonesians most wanted to visit in 2017, according to a survey conducted by Skyscanner (Kismadi, 2016). There are thousands of cosplayers in Indonesia, but only a small number of Indonesian cosplayers have visited Japan in order to do cosplay. Based on my interviews with nine Indonesian cosplayers, more often they attend cosplay events in neighbouring countries such as Singapore and Malaysia. Geographical proximity, the language barrier and financial issues are the main reasons.

> I am in love with Japan even though I have not been in there. I really want to visit Tokyo, Harajuku, Shibuya just like what I see in anime. You know the World Cosplay Summit, right? I want to cosplay in that event and meet a lot of foreign cosplayers. But, I am just a university student who does not have enough money to travel. It would be nice if I could do a part-time job, like what I see in *dorama* [Japanese dramas], but we do not have that kind of thing. It is not common for us and there aren't many part-time jobs available here. Maybe, after I graduate and get a steady job, I want save money to go to Japan. (Respondent S, 2018)

My biggest dream is to be able to cosplay in Japan, a sacred place for all cosplayers. But I am afraid because I cannot speak Japanese, although I want to communicate with Japanese cosplayers. I can only say a few words like *ohayō* (good morning), *arigatō* (thank you), *sukida* (I like you) etc. My friend said it will be challenging to visit Japan if I do not have knowledge of the Japanese language. Maybe I will get lost there if I travel alone [laugh]. Japan is quite far from here too ... So, for now, I prefer to go to Singapore or Malaysia to attend overseas cosplay events. The flight ticket and lodgings are affordable compared to Tokyo. My financial situation is tight because I have to spend money on my costume. (Respondent R, 2018)

Cosplayers who cannot attend overseas cosplay events owing to financial limitations can go to local cosplay festivals to feel the Japanese atmosphere. One of the most anticipated Japan festivals is Gelar Jepang Universitas Indonesia. GJUI is one of the oldest Japanese cultural festivals in Indonesia and has been held annually since 1994 (see Figure 9.3). Organized by the Japanology Student Association University of Indonesia (HIMAJA UI), this three-day event presents Japanese popular and traditional culture. Various activities are held such as *yukata* [summer kimono] fittings, a bon odori dance theatre, a manga exhibition and a cosplay competition, and the event ends with a fireworks show. From 2014 to 2018, GJUI was visited by around 26,000–30,000 visitors over the course of its three days (Nadya, 2015; Gelar Jepang Universitas Indonesia, 2017, 2018). The other main Japanese festival held in Indonesia is Ennichisai, which has been held since 2010. According to Kazeyuki, a hijab cosplayer from Bandung, Ennichisai has a good event concept so it is always attended by talented performers. There are many booths and Japanese-style decorations, such as cherry blossoms and food

Figure 9.3 GJUI 2018 visitors at the entrance of the venue (August 2018). Author's photo

stands. Fans of Japanese popular culture can feel the atmosphere of Japan without needing to go there.

There has also been an event organized by local government. On 2 November 2014, the Jakarta Capital City Government Tourism and Culture Office held the first Jakarta Cosplay Parade (JCP) in the National Monument (Monas). The event had the theme The Rise of Indonesian Legends. Attended by more than 500 cosplayers and 23 cosplay communities, JCP was the largest cosplay parade in Indonesia. The main characters were taken from Indonesian legends, such as Srikandi and Gatot Kaca. The aim was to increase awareness of these legendary characters among the younger generation. With this parade, it was expected that local characters would remain legendary despite being alongside popular characters from the USA and Japan. At a JCP press conference (Earlene, 2014), Basuki Tjahaja Purnama (the Governor of Jakarta 2014–2017), stated that the Jakarta Capital City Government supported the JCP because creativity has a significant contemporary role. Jakarta can be a domestic space for creative people, and JCP also helps make Jakarta into a tourism and cultural destination. In addition to the cosplay parade, JCP also held cosplay competitions and invited cosplay judges like Pinky Lu Xun. There were meet-and-greet events with international guest cosplayers, including Sharlene Liljaguar (Singapore) and Venus (Malaysia). The participation of international cosplayers was expected to make JCP an international cosplay event. As one JCP judge stated:

> The Jakarta Capital City Government promises that the JCP will be an annual event for Indonesian cosplayers to show their work. We want to have international events like Anime Festival Asia (AFA) and World Cosplay Summit (WCS) that can lift the national character. (Earlene, 2014)

Although it was planned to make JCP an annual event, there were no other events after 2014. This decision seems to have been influenced by the establishment of the Indonesian Agency for Creative Economy (BEKRAF) in 2015 (BEKRAF, n.d.). BEKRAF was established as a government effort to revive Indonesia's creative economy, such as animation and film. BEKRAF also supported the Indonesian animation *Si Unyil*, which was produced by the State Film Production Company (PPFN) and Telkom Indonesia in 2016. Si Unyil was originally a puppet film character who was popular in the 1980s and 1990s. The re-popularization of Indonesian characters such as Si Unyil and Si Bolang can be seen in their use in the introductory video at the World Cosplay Summit (WCS) by Ibun and Yumaki, Indonesia's representatives at World Cosplay Summit 2018

Indonesian Cosplayers' Behaviour Overseas

Cosplayers perform not only in national competitions, but also at the international level. One of the biggest cosplay competitions is at the WCS, which is held every year in Nagoya, Japan. The WCS was first held in 2003 and sponsored by TV Aichi Nagoya. Initially, there were only four participant countries: Japan, Germany, Italy and France. Later on, several countries including the USA joined too. In 2006, the WCS received Japanese government support from the Ministry of Foreign Affairs and the Ministry of Land, Infrastructure and Transport (now the Ministry of Land, Infrastructure, Transport and Tourism) (Suzuki, 2016). When Indonesia first joined the WCS in 2012, the Indonesian representative Rizki dan Yesaya won third place. In 2014, the Indonesian team of Guriinko and Ryan no Ryu won third place. Then in 2016, for the first time, the Indonesian competitors, Rian CYD and Frea Mai, became Grand Champions for their themed performance of *Trinity Blood*. Now, both of them are considered as the second generation of Indonesian cosplayer celebrities. The victory of Rian CYD and Frea Mai showed that Indonesian cosplayers are able to compete at the international level, even though cosplay is not originally from Indonesia.

> All this time, cosplay has always been underestimated by our society. Cosplayers are considered as people who have an identity crisis because they want to be anime characters. But since joining WCS in 2012 and 2014, Indonesian cosplayers have shown that we are able to excel at the international level. And in 2016 the Indonesian team was the champion. This shows that Indonesian cosplayers are able to make our nation proud. We have the skills to make costumes and perform at the international level. (Respondent M, 2018)

To determine how Indonesian cosplayers spend their time when attending cosplaying at events overseas, nine cosplayers were interviewed about their experiences. Frea Mai, the cosplayer from Jakarta who won the WCS in 2016, has also attended cosplay events in Singapore, Portugal and the USA. Her activities at the WCS were not just about the cosplay competitions. During her 12 days in Japan, she took part in the opening ceremony, a red carpet event, cosplay parade and cosplay judging. There were also tours designed by the committee for the representatives from all of the countries, such as visits to the theme park Meiji Mura, Laguna Ten Bosch and Inuyama Castle. Frea Mai prefers to be classified as an artist and feels less comfortable being called a celebrity cosplayer or an idol. For her, cosplay is not only about displaying her costumes, but also about her related work (for example, judging) and on-stage performances. Her testimony also reveals the demanding nature of her professional work as a cosplayer:

> The schedule was very tight. The organizer gave us one to two free days, but I was too tired, so I just slept. When I had time, I visited fabric and accessories stores in Nagoya. But there was nothing I needed, so I did not buy anything. Through WCS we build networks with international cosplayers. Rian and I were invited to be WCS cosplay judges in other countries such as Taiwan, Portugal and Spain. My cosplay experience abroad taught me a lot about costume design. It turns out that every

country has different cosplay tastes. For example, cosplay games are popular in Asia, while cosplay in Europe is more likely to be about an artistic character like a monster. (Frea Mai, 2018)

Two other cosplayer interviewees, R and M, also specifically looked for fabric and decoration stores after finishing their cosplaying trips in Singapore and Malaysia. Since some costume materials are not available in Indonesia, they try to find them when doing cosplay overseas. In addition, they said that attending cosplay events abroad helps them to get inspiration for tricks on how to make costumes in efficient and artistic ways.

Frea Mai also described how she enjoys socializing with fellow cosplayers at cosplay events. Through the WCS Japan network, she was invited by WCS organizers from other countries to become a cosplay judge and guest star. Even though she did not receive any honorarium when serving as a cosplay judge or guest star, the entire travel and accommodation expenses were provided by the organizer. In addition, the organizer also prepared a special booth for invited cosplayers to sell their merchandise, such as photo books and posters. Interestingly, Frea Mai claims to access her cellphone only rarely while attending cosplay events. As a result, she seldom takes selfies and prefers to focus on the event itself. If she has time, she prefers to take pictures of the scenery, the atmosphere of the event and the crowds. The difficulty of getting wi-fi was also one of the reasons she gives for infrequently updating her activities while abroad in real time. The issues of selfies also appeared in the interview with respondent W. She stated that she prefers taking pictures with other cosplayers while in costume. She could not even remember the location of pictures taken because she only focuses on the cosplayers.

> I love selfies, but I prefer to show my costume rather than being in pictures in front of the country's landmarks. I also enjoy taking pictures with other cosplayers. It was super fun and will be great memories for us. Sometimes I forget where I took some pictures because all I can see is just the cosplayers. (Respondent W, 2018)

Whereas some cosplayers have their travel expenses provided by organizers, some have to pay their own way. Rangga Kanchiel has attended several cosplay events abroad in Malaysia, South Korea and Australia specifically for cosplay and has paid for most of his own travel expenses. Respondent K and respondent W also stated how they participate in cosplay events specifically for cosplay as a passion that needs to be practised regularly. Respondent W attended cosplay events overseas almost every year from 2015 to 2018. Even though she did not take part in cosplay competitions, she was grateful when able to watch cosplayers' performances and get acquainted with cosplayers from other countries.

Rangga Kanchiel, who makes costumes and sells them in his online store, often promotes his products to fellow cosplayers. He stated that he was happy when his work got praised and was considered similar to the original costume. Besides promoting his merchandise, he also hunts for action figures that are not available in Indonesia. In his free time while abroad, instead of going to tourist attractions, Rangga Kanchiel prefers culinary tourism in the country.

Conclusion

Since cosplay was introduced to Indonesia in the late 1990s and early 2000s, cosplay has continued to grow and develop in Indonesia. Although cosplay is not a culture that originated from Indonesia, it began to be accepted within and adapted to Indonesian society and culture. Cosplayers do not only perform characters from the USA and Japan; Indonesian cosplay also shows that local characters can become icons and stimulate national pride. Cosplay has the power to transcend national borders, which shows that cosplay can be enjoyed by anyone.

Cosplay is generating new forms of tourism in Indonesia. In addition to the domestic and international tourism of Indonesian cosplayers, cosplay has also become a tourism attraction, such as in Bandung and Batu. The presence of cosplayers in Asia Afrika street, Cikapundung Terrace Park and Jatim Park received a good response from visitors, especially from children who could see their favourite anime characters. Regular cosplay events in Jakarta, such as GJUI and Ennichisai held by student associations, communities and the government, have made cosplay an alternative form of tourism. To ensure that cosplay generates sustainable tourism, cooperation between the government, the community and the private sector is needed. Organizing international cosplay events held on a regular basis attracts local and international tourists. Cosplay events also can be a marketing opportunity for business via the sale of costumes, merchandise and souvenirs to fellow cosplayers and visitors.

References

APJII (2010) *Penetrasi dan Perilaku Pengguna Internet Indonesia 2010*. Jakarta: Asosiasi Penyelenggara Jasa Internet Indonesia.
BEKRAF (n.d.) Badan ekonomi kreatif indonesia tonggak baru ekonomi kreatif indonesia. See www.bekraf.go.id/profil (accessed November 2018).
Beta, A.R. (2014) Hijabers: How young urban muslim women redefine themselves in Indonesia. *The International Communication Gazette* 4 (5), 377–389.
Bonnichsen, H. (2011) Cosplay-creating, or playing identities? An analysis of the role of cosplay in the minds of the fans. Unpublished MA thesis, Stockholm University.
Earlene, MM (2014) Jakarta cosplay parade press conference. See www.youtube.com/watch?v=OeOG-8IXQXg (accessed November 2018).
Gelar Jepang Universitas Indonesia (2017) Thank you for 30k+ visitors! See www.instagram.com/p/BXko1eoF265/ (accessed January 2019).
Gelar Jepang Universitas Indonesia (2018) Terima kasih kepada seluruh pengunjung, pengisi acara, sponsor, media partner, bazaar, dan seluruh pihak yang telah mendukung

penyelenggaraan gelar jepang universitas indonesia 24. See https://pbs.twimg.com/media/DmgErQ2VAAE8RLs.jpg (accessed January 2019).

Imanda, T. (2002) Komik Indonesia itu maju: Tantangan komikus underground indonesia. *Jurnal Antropologi Indonesia* 6 (2), 47–62.

Isadora, S. et al. (2012) Perbedaan motif afiliasi pada remaja pengguna facebook ditinjau dari jenis kelamin. *Intuisi Jurnal Psikologi Ilmiah* 4 (3), 1–5.

Jati, W.R. (2015) Islam populer sebagai pencarian identitas muslim kelas menengah indonesia. *Jurnal Teosofi* 5 (1), 139–163.

Kismadi, K. (2016) 5 destinasi yang paling ingin dikunjungi wisatawan indonesia di 2017. See www.skyscanner.co.id/berita/destinasi-yang-paling-ingin-dikunjungi-wisatawan-indonesia-di-2017 (accessed August 2018).

Lubis, I. (2009) Komik fotokopian indonesia 1998–2001. *Journal Visual Art & Design ITB* 3 (1), 57–78.

Nadya, F (2015) Press release: Gelar jepang universitas Indonesia 21 with wakuwaku Japan. See www.uiupdate.ui.ac.id/node/12754 (accessed September 2018).

Nuansya, A. (2017) Daya tarik wisata budaya festival cian cui di kota selat panjang provinsi riau. *JOM FISIP* 4 (2), 1–16.

Nugraha, F (2011) Sejarah dan masa depan perkembangan handphone berbasis android di indonesia. See www.teknojurnal.com/sejarah-dan-masa-depan-perkembangan-handphone-berbasis-android-di-indonesia/ (accessed November 2018).

Perdana, P.P (2016) Ridwan kamil promosikan bandung sebagai wisata cosplay. See www.nasional.tempo.co/read/786795/ridwan-kamil-promosikan-bandung-sebagai-wisata-cosplay/full&view=ok (accessed November 2018).

Pratama, H. (2018) Book review of Contents Tourism in Japan: Pilgrimages to 'sacred sites' of popular culture. *Jurnal Kajian Jepang* 2 (2), 114–117.

Rastati, R. (2012) Media dan identitas: Cultural imperialism jepang melalui cosplay. *Jurnal Komunikasi Indonesia Universitas Indonesia* 1 (2), 93–104.

Rastati, R. (2015) Dari soft power jepang hingga hijab cosplay. *Jurnal Masyarakat dan Budaya Lipi* 17 (3), 371–388.

Rastati, R. (2017) Hijab cosplay to hijab fashion industry in Indonesia. *Proceedings Asji International Symposium* 2017 (1), 211–221.

Reysen, S. et.al. (2018) Motivation of cosplayers to participate in the anime fandom. *The Phoenix Papers* 4 (1), 29–40.

Richa, I. (2018) Tokoh-tokoh anime jepang hadir ke the legend star, ada apa? See www.m.batutimes.com/baca/8860/20180127/201112/tokohtokoh-anime-jepang-hadir-ke-the-legend-star-ada-apa/?fbclid=IwAR3fXaVf2YDOxg6kXWp1Rde-LQkqTakSMNNgow-qMq2I8LiMWO7fMcl6Jdg (accessed November 2018).

Seaton, P. and Yamamura, T. (2015) Japanese popular culture and contents tourism – Introduction. *Japan Forum* (27) 1, 1–11.

Suzuki, R. (2016) Welcome to the cosplay olympics! Highlighting Japan. See http://dwl.gov-online.go.jp/video/cao/dl/public_html/gov/pdf/hlj/20160401/10-11.pdf (accessed November 2018).

Udasmoro, W. (ed.) (2017) *Dari Doing ke Undoing Gender: Teori dan Praktik dalam Kajian Feminisme*. Yogyakarta: Gadjah Mada University Press.

Interviews with cosplayers

(Interviewees who requested anonymity are represented with an initial only.)

Frea Mai, female, mid twenties, 6 May 2018, Jakarta
Kazeyukii, female, early twenties, 7 May 2017, Bandung
Makoto, male, mid twenties, 6 May 2017, Bandung
Rangga Kanchiel, male, late twenties, 6 May 2017, Bandung

Respondent K, male, mid twenties, 9 February 2018, Depok
Respondent M, male, early twenties, 13 February 2018 and 21 October 2018, Depok
Respondent R, male, late twenties, 11 February 2018, Depok
Respondent S, female, early twenties, 7 February 2018, Jakarta
Respondent W, female, late twenties, 10 February 2018, Jakarta

Part 3

Contents Tourism as Pilgrimage

10 Outbound Tourism Motivated by Domestic Films: Contentsized Koreanness in Thai Movies and Tourism to Korea

Sueun Kim

The transnational popularity of Korean popular culture, the Korean Wave, continues throughout Asia. Having started with TV dramas, now the Korean Wave has expanded to movies, variety shows, pop music and webtoons. The number of contents tourists visiting Korea is also increasing owing to the popularity of the Korean Wave. According to Chua and Iwabuchi (2008), the popularity of Korean pop culture – films, pop music and especially TV dramas – in the rest of East Asia became known as the 'Korean Wave' (Hallyu) by Chinese audiences in 1997, and the Korean Culture and Information Service (2011) identifies the popularity of Korean dramas broadcast in China in 1997 as the beginning of the Korean Wave. Based on this popularity, in 1999 the Korean government created an album of Korean pop songs called *Hallyu – Song From Korea*, which was the first official usage of the term 'Hallyu' (Jin & Yoon, 2017: 2244). Since then, Korean popular culture has gained popularity in Asian countries, and the term Hallyu has been used to refer to Korean popular culture consumed overseas.

Starting with the drama locations of *Winter Sonata* (2002) and *Dae Jang Geum* (2003), engaging with Korean entertainment contents has become one of the key motivations for Asian tourists to travel to South Korea. Thailand is one of the main countries in Southeast Asia where tourism related to Hallyu is gaining attention. Recently, the number of outbound Thai tourists to Korea has been rapidly increasing. The number of Thai tourists who visited Korea between 2007 and 2016 increased threefold. Affordable transportation (for example, Low Cost Carriers) and visa-free travel are major reasons for the increasing number of tourists, but as Jae-man Chon, Ambassador of the Republic of Korea to Thailand, has said, K-pop, Korean singers and dramas greatly influence tourism to Korea (Chon, 2014).

Furthermore, domestic Thai media contents that contain narratives related to Korea affect Thai tourism to Korea, too. This is not only because Korea appears as important locations, but also because 'Koreanness' appears in the narratives and characters. The Koreanness here is an imagined Koreanness created through media, in other words, a symbolic element formed in the perception of the audiences when experiencing Korean Wave products (Jeong *et al.*, 2017). Based on interviews with fans Ainslie (2016) has stated that 'perceived strong identity of "Koreanness" was a major source of popularity', so one may hypothesize that 'Koreanness' – whether in Korean or Thai productions – is a motivatational factor among those Thai fans who travel to South Korea as contents tourists. However, the Koreanness in domestic Thai media contents is 'contentsized' Koreanness in a sense. Takayoshi Yamamura defines contentsization as a 'the continual process of the development and expansion of the "narrative world" through both mediatized adaptation and tourism practice' in the introduction to this book. Koreanness in Thai contents is a reproduction of the narrative world derived from Korean contents. It is different from the actual image of Korea, but it reflects the ideas or perceptions held by Thai people of what the Korean people and culture should be and look like. This chapter examines how Thai media produces contentsized Koreanness, and how such Koreanness can attract Thai tourists to Korea through the example of the 2010 Thai movie *Hello Stranger*.

Koreanness in Korean Wave Contents

Before discussing Koreanness in Korean Wave contents, I will review how Koreanness itself was invented in the 1990s. Cho (1998) focuses on Koreanness as a process of finding new identities discussed in South Korea in the mid-1990s, just before the Korean Wave boom. The Koreanness at this time led to the Koreanness that appears later in Korean Wave contents. An increased interest in Korean tradition began with the government and businessmen, leading to a boom in Confucianism within the elite elderly and then to an interest in the tradition among the younger generation (Cho, 1998: 74). This was also a process of the production of hegemonic Koreanness, which was undertaken by governments, businesses and mass media within the context of globalization. To raise awareness of 'us', the national glory of the past was promoted, while 'others', such as colonial Japan, who are easily distinguishable from 'us', were identified. Confucian culture entered the spotlight again, and pop culture contents such as films based on Korean folk culture were made. Cho summarizes Koreanness in this period as 'a nationalistic attempt to regain national pride and capitalistic spirit in expanding into international markets' (Cho, 1998: 83–84).

Korean Wave contents inherited this Koreanness from the 2000s, both as a reality and an obsession. Koreanness, as a reality created in the 1990s, is said to be one of the factors that make Korean contents popular overseas. There has also been discourse about the overseas popularity of Korean Wave contents in which the existence of a certain unique Koreanness within the contents is suggested (Chung, 2013; Kim et al., 2009; Sung, 2010). Kim et al. (2009), for example, discuss the general Korean character appearing in Korean Wave contents as follows:

> (Korean) serial TV dramas focus on the domestic Korean audience, but are also distributed beyond national boundaries. Utilizing standard 'Western' production formats and technologies, they are written and produced by Korean production companies and cast Korean stars acting out Korean characters. They are generally set in Korean locations […] In other words, narratives, storylines, characters, actors, locations and music represented by the Korean TV dramas, whether set in the past or present, are explicitly Korean, condensing notions of Koreanness into an easily consumed popular format. Having said this, they tend to portray universal themes, including sudden sickness, accidents, love affairs, illegitimate children, envy, intrigue, betrayal and all manner of interpersonal entanglements. (Kim et al., 2009: 213)

A genre that can easily be said to exhibit Koreanness is historical dramas. Here the signals are obvious, including visual cues, such as costumes and sets, and narrative cues, such as plotlines set in an explicitly Korean past. A less clear-cut example is trendy dramas. Trendy dramas were influenced by Japanese dramas in the 1990s and have similar structures to those of other Asian countries/regions, such as China, Taiwan and Japan. However, overseas Korean Wave fans clearly distinguish between other dramas and Korean dramas. There are external elements which allow for this distinction, such as the Korean locations and Korean actors. There are also distinguishing elements of Korean dramas, such as specific emotions, sophisticated cosmopolitan characters, fantastic situations that suddenly appear in the real world, and ideal and romantic love. The important point is that Koreanness in the contents is producing a national image of Korea, and audiences and local media reproduce this Koreanness.

Such Koreanness is not only created within South Korean products. Shuling Huang presents an interesting study of how Koreanness is created through Taiwanese media. Huang analysed the process of the Taiwanese media's construction of Koreanness as follows:

> With the Taiwanese media's obsession with Japan and Korea, a sense of Japanese-ness and Korean-ness is reproduced in Taiwan – through media coverage, pop writings, TV drama productions and beyond. The local media's construction of Japanese-ness and Korean-ness is generic rather

than original. It derives from the Taiwanese emulation of Japan and Korea in accordance with their own imagination. Such desire for emulation, however, opens the door for the consumption of all things Japanese and Korean. (Huang, 2011: 12)

In other words, the Koreanness constructed by Taiwanese media is a reproduced and generic image drawn mainly from Korean media.

Furthermore, Yin and Liew (2005) analysed the acceptance of the Korean Wave in Singapore using the word 'imagined cosmopolitanism'. Yin and Liew emphasise that the Korean Wave in Singapore is part of Chineseness, rather than a direct acceptance and propagation of Koreanness by Chinese people, who account for about 75% of Singapore's population. This may have been valid in 2005, when the article was published, but since 2010, the Korean Wave has become its own form of cosmopolitan culture. One of the reasons for this is that the first Korean Wave in Singapore came from China. According to Yin and Liew (2005: 225):

When Korean serials were first aired in Singapore, they were obtained from Taiwan, Hong Kong, and China rather than from Korea. Korean music albums found in Singapore were repackaged in Singapore for a Chinese-speaking audience. Lyrics and song titles are translated into Mandarin.

However, by the beginning of the 2010s, Korean characteristics, that is, Korean identity, began to be transmitted directly to Singapore. K-pop went directly to Singapore without passing through China, and an internet community related to Korean pop culture was created (Lim *et al.*, 2016). This is closely related to the emergence of digital media, and similar phenomena occurred not only in Singapore but also across overall Korean Wave consumption. This direct and mutual Korean Wave consumption is called 'Korean Wave 2.0' and is distinct from the Korean Wave in the early 2000s (Lee S., 2015).

Contentsized Koreanness in Thailand

As Korean cultural products have gained in popularity, examples have emerged of pop culture industries in other countries/regions reproducing similar notions of Koreanness in local dramas and films. The example of Taiwan has already been mentioned. However, beginning with remakes of Korean works, many films, local musicals and dramas related to Korea have been produced since 2010 in Thailand, Indonesia, Malaysia and even Canada. Apart from the Canadian examples, most of these films and dramas were shot in and featured South Korea. The common feature of these works is that they represent both Korean tradition and the Koreanness seen in Korean Wave contents simultaneously in the narratives.

In Thailand, the most frequent pattern is a remake of Korean contents. One example is the Korean drama *Full House* (2004). Based on a 1993 Korean manga of the same title, the TV drama *Full House* spread across Asian countries and induced tourism to locations related to the drama in South Korea. The drama was remade in Thailand in 2014. The remake was made with the cooperation of the Korea Tourism Organization and half of the filming locations were in Korea. The remake generated synergy with the original and also renewed tourism relating to the original drama. The location of the 2004 drama was a house in Incheon, which was dismantled in 2013. In 2014, the Thai remake of the drama was filmed in Incheon on the invitation of Incheon City, and Incheon was able to promote tourist attractions relating to both the remake and the original.

The influence of Korean pop culture also makes some domestic Thai pop culture produced in Thailand more Korean in style. For example, in pop music, there are Thai pop singers (idols) who were trained in Korea, who then invite Korean choreographers to Thailand, or shoot music videos in Korea. The same goes for dramas. Along with the remake of *Full House*, several other Thai dramas were shot in Korea.

However, Koreanness produced through the media is not received in the same way by people across Thailand. Ainslie (2016) analysed how Koreanness is received by Thai people via Korean dramas. Ainslie concludes that there is a difference between urban and rural areas. In Bangkok, people tend to view Koreanness as creative and dynamic, as represented in the media, and regard Koreans in the dramas as role models for metropolitan Thais through shared identity as Asians. On the other hand, the rural people that Ainslie interviewed tend to perceive Koreanness simply as it is represented in the contents without considering the characters as someone to aspire to be (Ainslie, 2016). Although there are some differences in opinion among Thais regarding Koreanness, Ainslie concludes that the reception of Koreanness in Thailand is positive on balance:

> The construction of Koreanness clearly plays a part in enabling them to do this, indicating that both new forms of transnational cultural products and their increasing presence may play an important part in forming such national and international relationships. (Ainslie, 2016: 14)

The Koreanness that is being reproduced through Thai media spreads among Thai people and creates interest in travelling to Korea. As Jeong *et al.* (2017) argue, imagined Koreanness is created in the consumers' perception by consuming the Korean Wave. In that sense, Koreanness is an aggregation of individual elements related to Korea such as geography, history and culture through media representation. The Korean Wave plays a central role in creating images of Korea, increasing interest, and motivating Thai people to visit the country.

Some of these Southeast Asian travellers are disappointed by the gap between the Korean image created by the Korean Wave and the actual Korea they encounter on their travels (Teh & Goh, 2016). For example, in the Thailand Tourism Consumer Marketing Survey conducted by the Korea Tourism Organization in 2011, there is a review by a disappointed, anonymous Thai tourist who visited the theme park for the drama *Dae Jang Geum*. S/he was disappointed because there is nothing reminiscent of the drama in the theme park (KTO, 2011: 28). The *Dae Jang Geum* theme park is a shooting location built by Korean broadcasting company MBC. Various historical dramas have been filmed there, including *Dae Jang Geum*, which is a representative work of the Korean Wave. The park has become a famous contents tourism spot in Korea. The review does not go into great detail, but the tourists' disappointment may have been caused by the gap between the images in the contents and the set. *Dae Jang Geum* is about the life of an elegant woman who cooks at the palace during the Joseon Dynasty. However, in the theme park there is the set of the place where she cooks in the drama and a cheap-looking cardboard cut-out of the heroine. This tourist might have been expecting to see something more elegant.

Meanwhile, Khiun (2015: 123–126) argues that the movie *Hello Stranger* contributed to the formation of a new Koreanness and Thainess. According to Khiun, the image of Koreanness among Thai people created through Korean dramas also incorporates South Korean perceptions of Southeast Asia, especially of them as Third World people. Images of the Southeast Asians presented in Korean dramas are mostly those of migrant workers and marriage immigrants. Seeing such dramas, Thai people see South Korea as superior, or at least, they learn that Korean people see themselves as being superior to Thai people.

On the other hand, *Hello Stranger* portrays South Korea and Southeast Asia as being on the same level. The main Thai characters stay at a luxury hotel in Korea and enjoy gambling at a casino. Furthermore, sometimes the gaze is reversed and Koreans are presented as strange 'others'. In the film, for example, the male protagonist is surprised to see dog meat, and the two main characters eat raw octopus. Eating dog meat and raw octopus is part of traditional Korean cuisine, but this custom can elicit a critical reaction from people from other countries unused to the practice. At the same time, the South Korea depicted in *Hello Stranger* is an ideal space compared with Thailand. Khiun argues that South Korea in the film emerges as a free space that does not require any caution, either socially or emotionally. This is unlike Bangkok, where a coup took place in 2006. The two main characters of the film freely walk around the streets of South Korea while talking loudly, and wear the same clothes (the so-called 'couple look', in which wearing the same clothing in public is considered a part of 'couple culture' in Korea).

The Case of *Hello Stranger*

Hello Stranger (Thai title: *Kuan Muen Ho*, 2010) is a Thai romantic-comedy film produced by Banjong Pisanthanakun. It is the first movie that Thai moviemaker GTH created in collaboration with a Korean company (Korea Thailand Communication Company) and the whole film was shot in Korea (apart from the opening and closing scenes). The Thai moviemaker managed the scenario, casting and production, while the Korean side was responsible for finding shooting locations, hiring Korean actors and actresses, and the overall shooting. In addition, many Korean organizations, such as the City of Seoul, the Seoul Film Commission and the Korean Tourism Organization, supported the making of this film. The Seoul film commission provided $69,400 for the movie as a location incentive, gave the moviemakers permission to film in public areas, and provided helicopters to film scenes involving sweeping landscapes. The Korea Tourism Organization also promoted the movie (Lee N., 2015: 345).

The movie portrays a young Thai man, Dang, and woman, May, who meet by chance while on vacation on a package tour in South Korea. The man is a viewer of Korean contents, but is not interested in the Korean Wave. His motivation to travel Korea is not made clear in the movie, other than that he is forced by his friends to go on a package tour to Korea. Meanwhile, the woman is a big fan of the Korean Wave. Her motivation for travelling to Korea is to travel to the locations she saw in Korean dramas and movies. She is travelling by herself, but she has a boyfriend and in the first scene he is there to send her off at the airport. During their separate trips, the man and woman meet by chance and decide to tour Korea together while keeping their names secret from each other. The movie was filmed at various locations that have appeared in Korean dramas that were broadcast in Thailand.

The movie earned 4 million US dollars at the box office, making it Thailand's biggest hit of 2010 (Voicetv, 2010). The movie also garnered attention overseas. In Indonesia, the movie was the third biggest hit of the year, and it was also screened in Singapore and Malaysia (Sanook, 2010). The film was one of the rare Thai films to be shown in theatres in Australia. It was shown on three screens on 6 November 2010, but nevertheless all of the tickets for screenings of the movie were sold out (MThai, 2010).

However, *Hello Stranger* is not just a promotional movie for Korea. It received many Asian film awards. At the 2011 Osaka Asian Film Festival, it received the ABC Award for the most entertaining film, and it won the Most Promising Talent Award at the Suphannahong National Film Awards (Thairath, 2011a, 2011b). The film also received an award for Outstanding Performance by an Actress in a Leading Role at the 8th Starics Thai Film Awards, the popular vote at the 8th Komchadluek awards and the Outstanding Director award at the 2010 Top Awards, as well as the Asian New Talent Award at the 14th Shanghai International Film Festival (Voicetv, 2011).

In the film *Hello Stranger*, a reproduction of the Koreanness that appears in Korean contents and Thai people's ambivalence of this Koreanness are shown alongside each other. The main characters Dang and May visit places related to Korean dramas. As such, it is a film depicting contents tourism that has itself triggered contents tourism. In the film, drama locations in South Korea, such as Nami Island, which is a famous location from the drama *Winter Sonata* (2002), and the café in Seoul that was used as the location for the drama *The 1st Shop of Coffee Prince* (2007), are introduced as famous tourism sites in Korea. What is interesting is that the viewpoints of the protagonists regarding the Korean Wave are different. The heroine, a Korean Wave fan, visits a café that appeared in a Korean drama with sparkling eyes. On the other hand, the man travelled to Korea almost against his will. When he visits Nami Island he touches the face of a bronze statue of the drama star Yong-joon Bae thoughtlessly, mostly unaware or ignorant of the popularity of the Korean Wave. The behaviour of the man would be typical of Thai people with a negative perception of the Korean Wave, or those who are not Korean Wave fans but have many opportunities to watch because many TV channels broadcast Korean dramas, or who watch Korean dramas anyway before going to Korea. He went to Nami Island on a package tour, so he did not have much of a choice when it came to locations to visit. As mentioned above, it was not his choice to go to South Korea and in the movie's first scene the man's friends are seen sending him off to the airport. This made the tour relatively pleasant for him, but also a little uncomfortable.

At the same time, the movie's plot follows a pattern often seen in trendy Korean dramas, where coincidence overlaps and love deepens. For example, during their trip, the heroine May blacks out after having had too much to drink. May is surprised when she wakes up because she finds herself in the same room as Dang. Her mind immediately goes to clichéd scenes in trendy rom-com Korean dramas, and she is delighted but at the same time distressed because she has a boyfriend. As Kim (2005: 192) mentions, 'the trendy dramas of Korea, which invariably place romance and love at their center, are understood as reflecting the desires of women'. Going to a ski resort is another cliché often seen in Korean dramas. Such narratives describe the Korea-like situation that may occur while visiting Korea, and that Korea-like situations are nothing other than Koreanness drawn from Korean Wave contents. On the other hand, Dang says in the movie that 'Cheesy love stories like in [Korean] movies don't exist in real life'. This is a message to both the heroine and audiences in Thailand who like the Korean Wave. Paradoxically, or perhaps with deliberate irony, the protagonists have more dramatic experiences than characters in a Korean drama. Despite saying that it would be clichéd and impossible, Dang ends up having a fantastical experience, just as in Korean contents.

Meanwhile, fans see both Korea and Thailand through the film. While Koreanness is found in the narrative, Thainess is found in other elements. Pantip is one of the biggest internet communities in Thailand. Created in 1997 and with 45 million users per month, it is the fourth largest site in Thailand after Google, Facebook and YouTube. Some of the fan comments on Pantip include:

- *Hello Stranger* can be said to be a legend among GTH's productions. At that time, it was a new challenge, and the main characters were also good. It is a fun and impressive movie.
- Personally, I think *Hello Stranger* is a unique movie. It is the pride of Thai movies. I was not surprised at the success of this film, and there are many overseas fans of this film.
- Because most of the public like feel-good or happy-ending movie genres, I think *Hello Stranger* fits the public's expectations.
- Like sweet coffee, *Hello Stranger* is easy to consume for everyone (Pantip, 2016).

This film has induced Thai people's tourism to Korea. For example, a post titled, 'Suddenly I wanted to travel to Korea after watching *Hello Stranger*', was posted in the Thai community site Pantip on 23 August 2010 (Pantip, 2010). There were also actual travellers' responses such as, 'I went to Korea because of *Hello Stranger*', and 'It's a start. The movie makes us want to travel'.

It is difficult to quantify the exact impact of the film in numerical terms (either numbers of visitors or economic impacts), but the film has clearly had a significant influence on Thai tourism in Korea. As in many examples of contents tourism, it is difficult to separate out the direct effect of contents on tourism from other social, cultural and economic factors affecting tourism. The film came out at a time when there was an overall increase in Thai tourists to Korea (Figure 10.1).

Furthermore, KTCC, the Korean co-production company of *Hello Stranger* posted the following anecdotal evidence about the tourism induced by the movie on their website.

> The KTCC received an appreciation plaque from the Seoul Metropolitan Government. Ji-hee Hong, CEO of KTCC, received a plaque of appreciation from Seoul Mayor Se-hoon Oh for the increase in the number of tourists in Seoul at the 7th floor executive meeting room of the Seoul City Hall annex, 16 August 2011.

> The number of Thai tourists who are heading for Seoul has sharply increased due to the movie *Hello Stranger.* On this day, the director of the film, the main actor and actress, and the Thai producer at GTH were also presented with a plaque of appreciation.

Mayor Se-hoon Oh said, '*Hello Stranger* was an opportunity to inform the people of Southeast Asia of Seoul's beauty. The number of Thai tourists heading for Seoul, which is one of the locations of the film, increased by 40% compared to last year. The movie greatly contributed to Seoul's tourism industry'.

In conjunction with the film, the 'Seoul Tour' product which takes tourists to the locations of the movie – Deoksu Palace, Namsan and Myeongdong – was developed. Last year, about 3,000 Thai tourists participated in this tour. (KTCC, 2011)

A characteristic feature of *Hello Stranger*-related travel is that travellers are taking package tours as well as making private trips. Much of contents tourism is based on individual travel, particularly when fans are on a pilgrimage, such as an 'anime pilgrimage'. However, the story in the film is about a romance during a package tour, so in this case package tours seem to be an appropriate mode of contents tourism. People who have been on these package tours have also left accounts of their travels, such as the blog post titled 'Snow play in Korea – Following *Hello Stranger* and going to Nami Island', which was posted on a personal blog on 14 February 2011 (OK Nation Blog, 2011). The writer travelled to South Korea on a package tour, visited Nami Island, posed in the same way as in a specific movie scene and introduced South Korea as the location of the movie *Hello Stranger*. In other words, contents tourism itself becomes part of the narrative world and *Hello Stranger* offers a good example of contentsization.

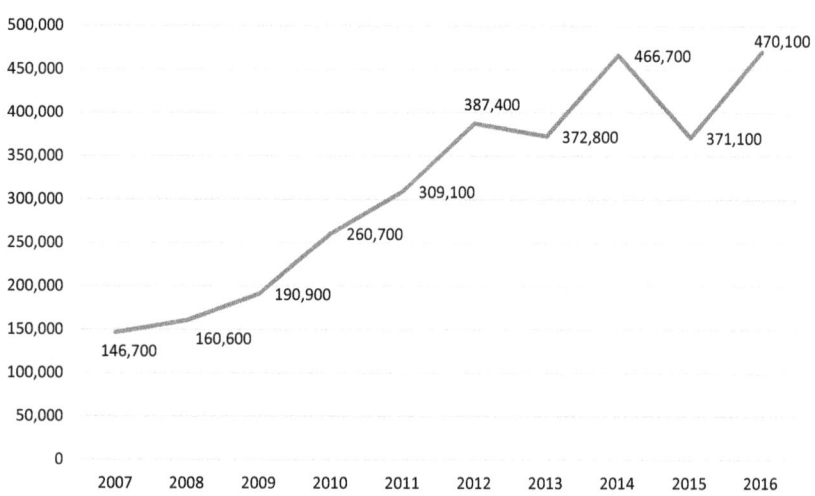

Figure 10.1 Number of Thai tourists to Korea between 2007 and 2016

Conclusion

This chapter has examined how domestic Thai films entice Thai people to travel to Korea, focusing on Koreanness as a contentsized destination image. Koreanness as depicted in Thai media contents is the reproduced and generic image drawn mainly from the Korean media. In other words, what is expressed in one imagination space is re-expressed through another imagination space. Such Koreanness in Thai contents is interpreted according to the tastes and preferences of Thai people, and they appeal to the audiences of Thailand more easily than the purely Korean contents.

The movie *Hello Stranger* is a good example of the expression of contentsized Koreanness and a work that has generated a rise in tourism. The movie shows how Koreanness is reproduced through local contents. While it may be stereotypical for Koreans to be simplified and romanticized in this way, it is a representation of South Korea that Thai people actually see and feel through the media. The two main characters' experiences in the movie may become the type of experience that Thai people long for and imagine should they ever have the chance to visit South Korea themselves.

On the other hand, the Koreanness in the contents is different from 'actual' Koreanness. Thai people experience the gap between the actual South Korea and the Korea within the contents when they visit Korea. Based on interviews with 80 Southeast Asian Korean Wave consumers, Ainslie *et al.* (2017: 77) note that 'Due to this "gap" between the real and the fantasy [...] the clean-cut fantasy projected in Hallyu products cannot match the actual reality of Korean society, with consumers noting their disappointment'. Thus, this gap between reality and fantasy can also change fans' perception of South Korea negatively.

Nevertheless, such a gap cannot be said to be exclusively negative, as it provides an opportunity to the fans to consider Koreanness amidst the ongoing popularity of the Korean Wave. Such a process is an opportunity for Thai people to understand South Korea more deeply. In these ways, the meanings of contentsized Koreanness and related tourism contribute to the discovery the 'real South Korea'.

References

Ainslie, M.J. (2016) K-dramas across Thailand: Constructions of Koreanness and Thainess by contemporary Thai consumers. *The Asia-Pacific Journal* 14 (7), 1–15.

Ainslie, M.J., Lipura, S.D. and Lim, J. (2017) Understanding the potential for a Hallyu 'backlash' in Southeast Asia: A case study of consumers in Thailand, Malaysia and Philippines. *Kritika Kultura* 28, 63–91.

Cho, H. (1998) Constructing and deconstructing 'Koreanness'. In D.C. Gladney (ed.) *Making Majorities: Constituting the Nation in Japan, Korea, China, Malaysia, Fiji, Turkey, and the United States* (pp. 73–91). Redwood City, CA: Stanford University Press.

Chon, J. (2014) Blood pledge Thailand, must maintain a good image of Korea. *Chindia Plus* 95, 18–19.

Chua, B.H. and Iwabuchi, K. (2008) Introduction: East Asian TV dramas: Identifications, sentiments and effects. In B.H. Chua and K. Iwabuchi (eds) *East Asian Pop Culture: Analysing the Korean Wave* (pp. 1–12). Hong Kong: Hong Kong University Press.

Chung, P. (2013) Co-creating Korean Wave in Southeast Asia: Digital convergence and Asia's media regionalization. *Journal of Creative Communications* 8 (2–3), 193–208.

Huang, S. (2011) Nation-branding and transnational consumption: Japan-mania and the Korean wave in Taiwan. *Media, Culture & Society* 33 (1), 3–18.

Jeong, J.S., Lee, S.H. and Lee, S.G. (2017) When Indonesians routinely consume Korean pop culture: Revisiting Jakartan fans of the Korean drama *Dae Jang Geum*. *International Journal of Communication* 11, 2288–2307.

Jin, D.Y. and Yoon, T.J. (2017) The Korean Wave: Retrospect and prospect. *International Journal of Communication* 11, 2241–2249.

Khiun, L.K. (2015) Into the heart of the Korean Wave in Banjong Pisanthanakun's *Hello Stranger* and Poj Arnon's *Sorry, Sarangheyo*. In M.J. Ainslie and J.B.Y. Lim (eds) *The Korean Wave in Southeast Asia: Consumption and Cultural Production* (pp. 115–132). Petaling Jaya: Strategic Information and Research Development Centre.

Kim, H., M. (2005) Korean TV dramas in Taiwan: With an emphasis on the localization process. *Korea Journal* 45 (4), 183–205.

Kim, S., Long, P. and Robinson, M. (2009) Small screen, big tourism: The role of popular Korean television dramas in South Korean tourism. *Tourism Geographies*. 11 (3), 308–333.

Korean Culture and Information Service (2011) *The Korean Wave: A New Pop Culture Phenomenon*. Sejong-si: Korean Culture and Information Service, Ministry of Culture, Sports and Tourism.

KTCC (2011) Seoulsi, taegug-yeonghwa 'hello seuteuleinjeo' e gamsapae suyeo. See http://www.k-tcc.com/page/sub72.php?boardid=JS_board_board07&mode=view&no=117&start=80&search_str=&val=&sort= (accessed October 2019).

KTO (2011) Thailand Tourism Consumer Marketing Survey. See http://kto.visitkorea.or.kr/viewer/view.kto?id=32804&type=bd (accessed September 2018).

Lee, N. (2015) Analysis of screen tourism and location strategies in the Thailand cinema <Hello Stranger>, *Journal of Communication Design* 53, 338–349.

Lee, S. (2015) Introduction. A decade of Hallyu scholarship: Toward a new direction in Hallyu 2.0. In S. Lee and A.M. Nornes (eds) *Hallyu 2.0 The Korean Wave in the Age of Social Media*. Ann Arbor, MI: University of Michigan Press.

Lim, T.W., Lim, W.X. and Ping, X. (2016) Korean Wave (Hallyu) in Singapore: Policy implications of *Hallyu* and its background – The Southeast Asian context. In T.W. Lim, W.X. Lim, X. Ping and H.Y. Tseng (eds) *Globalization, Consumption and Popular Culture in East Asia* (pp. 183–196). Singapore: World Scientific.

MThai (2010) Kuan muen ho dangklaitangdaen ottrelia temthukrop triamchokhiochaiiklaiprathet. See https://movie.mthai.com/movie-news/81926.html (accessed September 2018).

OK Nation Blog (2011) Luihimakhaothikaolitai ko nami tamroi kuan muen ho. See http://oknation.nationtv.tv/blog/storyman/2011/02/14/entry-1 (accessed June 2018).

Pantip (2010) Dukwon muen ho ma ayakpaikaolimakkhrapnaenamthi. See http://topicstock.pantip.com/blueplanet/topicstock/2010/08/E9609479/E9609479.html (accessed June 2018).

Pantip (2016) Kuanmuenho kap faende khunchoprueangnaimakkwakankhrap. See https://pantip.com/topic/35872480 (accessed June 2018).

Sanook (2010) Kuan muen ho dangklaithuengtangdaen titandap bokoffitindo. See https://www.sanook.com/movie/16733/ (accessed September 2018).

Sung, S-Y. (2010) Constructing a new image: Hallyu in Taiwan. *European Journal of East Asian Studies* 9 (1), 25–45.

Teh, P.Y. and Goh, H.C. (2016) Does Korean drama have a real influence? An analysis of Malaysia outbound tourists to South Korea. *Tourism Culture & Communication* 16 (3), 147–160.

Thairath (2011a) Kuanmuenho khwarangwan thetkannangthiosaka. See https://www.thairath.co.th/content/155739 (accessed September 2018).

Thairath (2011b) Nunaananda sionamchaiying supronnanga 53. See https://www.thairath.co.th/content/153933 (accessed September 2018).

Voicetv (2010) 10 andapnangthai thamraidaisungsutnaipi 2010. See https://www.voicetv.co.th/read/1345 (accessed September 2018).

Voicetv (2011) Toe khwarangwandaradaorungchaihaengechia. See https://www.voicetv.co.th/read/13141 (accessed September 2018).

Yin, K.F.S. and Liew, K.K. (2005). Hallyu in Singapore: Korean cosmopolitanism or the consumption of Chineseness? *Korea Journal* 45 (4), 206–232.

11 Contents Tourism in Plane Sight

Christopher P. Hood

> Hello ladies and gentlemen who love STAR WARS. This is the Captain speaking. On behalf of Han Solo, I will be in charge of this jet today. For your safety, hyperdrive will not be used for this flight.
> (ANA, 2015b)

Contents tourism primarily involves fans travelling to places related to a particular element of popular culture of which they are a fan. That is the core of the definition presented by Seaton *et al.* (2017: 3). While this volume presents a modified definition, the focus of the majority of the chapters contained within it is on activities happening at a destination, with the chapter by Beeton being one of the few that points to other possibilities. This chapter aims to ensure that the way in which we think about the sites of contents tourism does not become overly restrictive. Furthermore, the chapter raises questions about where the line is drawn between consumers, fans, tourists and commercial operations, which, at least on the surface, appear to be trying to profit from contents tourism activities. To do this, the chapter considers contents tourism in relation to planes, which, as the title suggests through a play on words with the phrase 'in plain sight' can seemingly happen with some ease.

Just as contents tourism has been primarily concerned with fans travelling to a place, so travelling by inter-city transportation, such as planes and high-speed railways, is for most passengers about getting to a destination. Rare are those who go to the expense of purchasing a ticket because they just want to experience the mode of transportation itself and who would, presumably, be looking to return to their home soon or for whom the destination is of lesser interest than the time travelling. Yet this chapter will show that contents tourism can involve the mode of transportation. This chapter will focus on primarily on planes owing to their potential international reach as this volume is seeking to highlight the transnational nature of contents tourism. However, in so doing the chapter also questions what we understand by 'national', without which 'transnational' cannot exist.

Although there have been TV dramas and movies set on planes, some using real company names, the planes themselves in such dramatization tend not to have any distinguishing features, and the uniforms of the crew, for example, would be the same or similar to those used by the real companies. This makes it hard to identify passengers who may be using a particular airline owing, in part or in totality, for contents tourism reasons. Rather, this chapter focusses on contents tourism in relation to the liveries on the outside of the planes.

To address the aims of this chapter, the first section will present some context in relation to the role of liveries and Japan's 'wrapping culture'. In the second section, background information about the Japanese airline industry and the liveries of Japanese planes is presented. This will be followed by introducing examples of popular culture found on some Japanese planes. The final section will pull together the key findings of the chapter, showing how these liveries should be considered as contents tourism, and in so doing, pointing how these both aid and challenge our understanding of what contents tourism is, where it happens and even who the contents tourists are.

This chapter is based on extensive research about transportation in Japan conducted over an 18-year period and which has included using all of the Japanese airlines discussed within the chapter, visits to and interviews at airline companies, as well as observations at and near airports.

Plane Liveries and 'Wrapping Culture'

Before we discuss the sorts of liveries used by Japanese airlines, we need to consider what the role of a livery is. Despite the fact that airline travel has become a more and more common part of life, as Lee *et al.* (2018: 533) and Wang (2015: 100) note, it is an area that has not received much academic attention. While Budd (2012: 63–64) notes that liveries began as rather simple designs and the function was primarily to aid in protecting the plane from external damage and to aid with its aerodynamic features, over time the designs became more intricate. Consequently, they are ripe for study of their symbolic meanings (Thurlow & Aiello, 2007).

When looking at airlines around the world, their establishment, development and growth have often stemmed from imperial, colonial and diplomatic links. While not all companies necessarily set out to become a national symbol or be seen as the nation's main airline, the drive to compete, survive or grow can lead to the domestic/national becoming more international/transnational. Consequently, it is hardly surprising that airlines have a long history as being seen as national symbols (Raguraman, 1997: 240), and an airline may even carry 'national flag carrier' status, which often means that it has certain privileges from the national government. Yet, as with the case of PanAm, being seen as a national symbol may not always be positive. It ultimately contributed to PanAm's demise as the carrier became a focus of international terrorism. Thurlow and Aiello (2007) also note the problems that British Airways faced when it decided to drop the national flag from its livery, how this led to financial problems and the how the airline eventually had to go back to using the national flag.

Although Budd (2012) considers the whole livery, Thurlow and Aiello (2007) and Lee *et al.* (2018) focus their attention purely upon the design of the tail fin. While the greater whole-plane approach of Budd has key advantages owing to it not overlooking the message that can be understood from the total design, and these three studies establish certain key features that are used by airlines, all three studies suffer from one key problem: they work on the assumption that all planes by a single carrier are identical. In fact, as this study will highlight, airlines may have a variety of different designs. That this is the case is also highlighted by Wang (2015) and Wang and Ngamsiriudom (2015) in their studies of special liveried planes.

Whilst all of the studies mentioned thus far refer to the role of branding in livery designs, I would like to present another conceptual way to think of the liveries. Hendry (1993) has written about the importance of the 'wrapping culture' and, while noting that it exists in many cultures, that it appears to be particularly significant in Japan. Indeed Hendry (1993: 172) suggests that in Japan the external appearance, the 'wrapping' is 'a veritable "cultural template" or [...] a "cultural design"'. Examples of this 'wrapping' can relate to how the wrapping of a gift can be just as critical as the contents and the 'wrapping' that occurs within the Japanese language through the use of a variety of levels of politeness (Hendry, 1993). Doi (1989: 25–26) also suggests that through the 'wrapping', or *omote* as he describes it, the Japanese also understand what lies behind it. I find these ways of understanding the outside appearance of things very compelling and have previously used this in seeking to understand the design of the *shinkansen* ('bullet train') (Hood, 2006: 155–157). Similarly, the 'wrapping' of the plane presents a message about the airline that operates the plane, as will be discussed in this chapter in relation to contents tourism.

Attention to the livery raises the question as to who sees it, and by extension, who it is for. There appear to be two main groups. The first is the flyers – the passengers who take the plane. While the nature of the flyers can be very diverse in terms of their motivations for taking a particular flight, they can be broadly divided into business and vacation travellers. Although some airlines may specialize in providing services to one type of flyer, and the brand image may speak to that group more than the other, in reality many flights may have a combination of both groups represented (Lee *et al.*, 2018).

Although the plane's livery is not the only place where a company's brand is on display, it is arguably the most iconic since it is the most public facing and so closely related to the core business. Yet with many airports providing air bridges that take passengers directly to the plane, the opportunity to see the plane they are about to board can be limited. Furthermore, with planes lined up perpendicular to the terminal buildings, the air bridges can also obstruct the view of the livery along the side of the plane. Once planes have taken off, they cruise at an altitude such that they remain largely out of sight in a way that does not happen with trains, for example, which are often visible from line-side vantage points. However, while taxiing, during landing and take-off, the planes may be visible. Unlike the situation in many countries, all Japanese airports have observation decks and there are often parks and other places near the airport where people can go to take photographs. Consequently, the second group who take an interest in the plane livery is photographers. Planes with special liveries are very popular with this group, and many I have interviewed have commented on how they will check the schedule of planes, or flight-tracking apps on their mobile phone, to ensure that they can see such planes. Photographs of these planes may then appear on social media, acting as a form of free advertising for the airline. However, understanding the motivations of photographers is not easy. In relation to the focus of this volume on contents tourism, would a photograph of a plane with popular culture characters on it, for example, only be contents tourism if an individual primarily takes pictures related to that particular aspect of popular culture? If an individual often takes pictures of planes, is it something different to contents tourism when they take such a photograph?

Although flyers and the photographers appear to be the two key groups, this chapter will ultimately argue that the main contents tourism is being done by another actor.

Japanese Planes

When discussing transportation in relation to Japan, the *shinkansen* is internationally renowned and the image of the *shinkansen* passing Mount Fuji has become synonymous with Japan (Hood, 2006). However, high-speed railways are not the only way to get around Japan. Japan has a developed airline network too, with 97 airports dotted across the country (MLIT, 2016). Although there are aspects of contents tourism which should be studied by looking at the names and activities at some of these airports, this chapter will focus on the Japanese airlines which serve them, and leave airports for future research.

When it comes to the Japanese airline companies, Japan Airlines (JAL) is still seen by many as the national flag carrier, although it no longer has the largest market share in Japan (CAPA, 2018). Throughout its history the JAL logo (also used by other airlines, such as JAL Express, in the JAL group) has had a national symbolic quality. Its original logo, known as the *tsurumaru*, a version of which was brought back in 2011, uses a red crane in a circular shape, linking it to the Hinomaru national flag (JAL interview, 25 July 2013). Its other logo, used on the tail fin, was even more symbolic in its relationship to the national flag as it was part of a red disc. The degree to which studies such as Thurlow and Aiello (2007) pick up on all of the symbolic meanings of such designs is questionable. Although most JAL planes use a standard livery, some use a livery to show its link to an international alliance, the OneWorld Alliance, or to aid with promoting particular campaigns, such as trying to encourage people to travel to a particular area of Japan (see Figure 11.1).

Whilst JAL may be the airline that many will associate as being Japan's main carrier, All Nippon Airways (ANA), which started out as a helicopter company, hence its NH (Nippon Helicopter) flight code (ANA interview, 23 July 2013), is the largest Japanese airline now (CAPA, 2018). ANA was originally limited to domestic services but has grown its international network in recent years following restrictions being lifted by the government (MLIT interview, 9 August 2013). In terms of its liveries, on most planes 'ANA' appears with the airline's blue logo (in contrast to JAL's use of red), which also forms the basis of the colour scheme for the tail fin on the majority of its planes. However, ANA has found from some of its international advertising campaigns that, after it promotes reasons to go to Japan, potential customers then book with *Japan* Airlines rather than ANA, as the company's name does not have the same obvious name link with Japan

Figure 11.1 (a) A JAL plane in the standard *tsurumaru* livery, (b) a JAL plane sporting a 'Let's visit Kyushu' campaign livery, (c) an ANA jet with the additional 'Inspiration of Japan' slogan on it and (d) The Panda ANA jet. Author's photos

(ANA interview, 23 July 2013). Consequently, many of the planes now feature the phrase 'Inspiration of Japan' written down the fuselage, using the company's slogan which was adopted in 2010 (ANA interview, 23 July 2013). The use of the word 'Japan' not only ensures that far more people will know where the plane comes from, but it also aids with educating people so that they accept the 'ANA' symbol as a being that of a Japanese airline, even if they do not know what the letters stand for. Whilst the current logo appears on most of their planes and many planes will be in the company colours, ANA uses a much greater range of liveries than is seen with JAL planes. The designs have included planes painted to resemble a blue whale, a panda design, those which show its membership of the international Star Alliance, and those which are related to popular culture, as will be discussed in the following section (see Figure 11.1).

In addition to the two legacy carriers, JAL and ANA, there are a number of smaller hybrid carriers and Low Cost Carriers (LCC) in Japan,[1] many of which are owned in part or otherwise affiliated with JAL or ANA. While Solaseed Air operated under its previous name of Skynet Asia Airways, planes had brightly coloured flowers on them to resemble those found in the tropical areas of Japan which the company primarily served. However, following its rebranding in 2011, the planes became predominantly green and white with the new Solaseed logo featuring on the tail fin and the company's name down the fuselage (Solaseed interview, 5 September 2013). Many of the planes also carry a motif of local villages, towns or cities on the island of Kyushu, which the company serves. At the other end of the country, Air Do is based in Hokkaido and links cities across that island with Tokyo and a few other cities. Whilst most of its planes are based on a livery using its sky-blue and yellow corporate colours, some planes also include cartoon pictures of bears, an animal associated with Hokkaido. In both these cases, the airlines are using the liveries as a means to provide a link to their regional identity (Air Do interview, 26 August 2013).

When considering the liveries of the other hybrid airlines (Skymark, Starflyer, Ibex and Fuji Dream Airlines (FDA)) and LCC airlines (Jetstar Japan, Peach, Spring Airlines Japan, Vanilla Air), the main feature which stands out is a consistent lack of any of the regional designs found on Solaseed and Air Do planes. Whilst one may expect Starflyer, which is also a regionally based hybrid (based in northern Kyushu), to play up that regional identity, its distinctive black livery has been at the centre of its brand, which also emphasizes a level of luxury that is often not found on other hybrid carriers or LCCs (Starflyer interview, 3 September 2013). Most airlines want to have a consistent livery and design to help raise brand awareness (Peach interview, 5 August 2013; Jetstar Japan, interview 6 September 2013; Vanilla Air interview, 15 May 2015; Spring Airlines Japan interview, 15 May 2015). In this respect, FDA is unusual in that each of its planes is painted a different colour.

Looking at the planes that have special liveries discussed thus far, one feature that stands out about many of them is the use of 'cute' characters and designs in many cases. Kinsella (1995) notes that cute (*kawaii*) culture is something that companies have been exploiting since around the 1970s. Indeed, it has become so pervasive that Kageyama (2006) argues that it is 'gaining such overseas acceptance it's rapidly becoming Japan's global image'. Indeed, *kawaii* was a cornerstone of the 'Cool Japan' concept, which was adopted by the Japanese government in trying to promote the country internationally (Craig, 2017). Although none of the airlines that I interviewed said that their designs were a part of 'Cool Japan' itself, there can be little doubt that the designs would fit with what many foreign visitors would expect to see in a country where *kawaii* has become such a feature. However, whilst there is something 'Japanese' about the designs, the use of a panda design, for example, shows that that the airlines do not restrict themselves to things found in Japan. Furthermore, arguably Japan's most well-known *kawaii* icon, Hello Kitty, has not featured on a Japanese plane livery, but has appeared on planes of Taiwan-based Eva airlines flying to Japan (Wang, 2015). However, it is important to stress that most planes in Japan do not have a special livery and so do not have any characters, let alone cute ones, on them. Indeed, with the exception of JAL and other airlines in its group or Spring Airlines Japan, which also have the name 'Japan' in them, it would be difficult to identify most of the planes as being Japanese at all. Although plane registrations start 'JA' and planes may have a small Japanese flag on them, to the casual observer, particularly if seeing a photograph of a plane without knowing where the photograph was taken, the national identity is likely to be unclear. This apparent lack of nationality, at least in the 'wrapping' of the plane, is significant in aiding us to understand some of the popular culture that appears on Japanese planes and for then considering contents tourism in relation to this.

Japanese Planes and Popular Culture

As already discussed, a number of Japanese airlines use non-standard liveries on some of their planes. In addition to the types which have already been introduced, some airlines use liveries that have a more direct connection to popular culture. This section will provide examples of some of these, before the next section considers why such liveries are used and why it is appropriate to consider them within the concept of contents tourism. In this respect it is important to note that planes are not the only transportation that may have characters related to popular culture on them in Japan. Seaton *et al.* (2017: 41–42) show how they may appear on trains and *itasha* (private cars decorated with scenes from popular culture) and characters such as Pokémon, Evangelion and Hello Kitty have been used on the *shinkansen* (see Figure 11.2).

Figure 11.2 (a) A display in the second carriage of the Hello Kitty *shinkansen*; (b) logos on the outside of the train featuring two of the prefectures that the Hello Kitty *shinkansen* stops at; and (c) the Hello Kitty *shinkansen* at Mihara station. Author's photos

Pop culture appears on plane liveries in two different ways. First, the design is limited to a section towards the back of the fuselage, mostly behind the main wings. Examples of this type have included JAL's use of the image of members of the pop group Arashi (see Figure 11.3a), Japan Transocean Air (a subsidiary of JAL) using the image of singer Namie Amuro on one of their jets to mark her retirement from the music industry in 2018, and JAL using Disney characters on a plane as part of campaign to celebrate an anniversary since Tokyo Disneyland opened and to encourage people to visit the resort. Second, in other cases, much of the plane may be covered with the imagery, as ANA did with its Pokémon jets (see Figure 11.3b). In relation to the discussion of national/transnational elements and the fact that the national identity of many Japanese planes may not be obvious, although Arashi and Amuro are known beyond Japan, their national identity is clear. On the other hand, the use of Disney characters presents a more interesting dynamic as the theme park, Tokyo Disneyland, which they are being used to promote, is a Japanese resort, but the characters themselves are American, albeit with a very global reach. In other words, even on planes which are going to be used domestically, non-Japanese characters, even if linked to a national tourist attraction, can be used.

Figure 11.3 (a) A JAL plane with pop group Arashi on the livery; and (b) One of the ANA Pokémon 747s. Author's photos

Whereas the origins of Disney are in the USA, Pokémon are Japanese and so at one level it may seem less surprising to find them on a Japanese plane. However, we need to keep in mind the Disney characters were there to promote a destination, Tokyo Disneyland, not the characters themselves. This is different to what was seen with ANA's Pokémon jets – there was no Pokémon 'land' to promote and the last of the five planes, which had been operating since 1998, retired in 2016, before the Augmented Reality app, Pokémon Go, which encouraged people to use their mobile phone and travel around 'catching' Pokémon, gained popularity.

However, whilst ANA has stopped its association with one well-known popular culture brand, the airline has adopted arguably an even more well-known film franchise in the form of Star Wars (see Figure 11.4). In addition to the planes themselves, ANA has posted a number of videos related to its Star Wars jets on its 'ANA Global Channel' on YouTube (e.g. ANA, 2015a, 2015b, 2016) and there is a dedicated website (ANA, 2018). Looking through these various media, one could get the impression that the planes are a haven for contents tourists. One video, for example, shows some fans wearing Star Wars costumes for a flight (ANA, 2015b). However, it would appear that most flights do not see such levels of interaction – searching the Internet it is difficult to come across evidence, such as photographs, of people dressing up as Star Wars characters when taking one of these planes, other than those who did so

Figure 11.4 (a) The ANA C3-PO Star Wars jet; and (b) the ANA BB-8 ANA Star Wars jet. Author's photos

for the flight which featured in the ANA video. Indeed, although Star Wars planes have Star Wars-related items on the inside (see ANA, 2017) and ANA has a range of Star Wars plane merchandise, I have come across passengers who have flown on Star Wars jets without even being aware that they were on one, having not been able to see the plane clearly before boarding. On the other hand, another interviewee commented on how the ANA pilot of the plane they were on pointed out the 'special' plane taxiing past them in the opposite direction, which passengers by the window might want to see, as one of the Star Wars jets went by. However, there is no obvious connection between Star Wars and Japan, in the way that exists, for example, with Japan and Hello Kitty, which EVA Air has on some of its planes and which Wang (2015) found was an effective livery owing to the link between Hello Kitty and the destination. Although some may go to the Star Wars rides at Tokyo Disneyland, and it could be argued that ANA, in using robot characters rather than any of the human-like ones, may be making a link between Japan and its image of a high-tech robot-using country, the link is clearly more tenuous than that of Hello Kitty and Japan.

Therefore, we are left with a bit of puzzle. Can these planes which display icons of popular culture on them be understood as being a part of contents tourism and, if it is, who is doing the contents tourism?

Understanding the Contents Tourism of Japanese Planes

Despite the lavish liveries, ANA say that they have not done any particular cost–benefit analysis as to whether these liveries help bring in customers (ANA interview, 23 July 2013). So why do they do it? As noted above, ANA needs to do things to raise its international awareness. Compared with the Pokémon planes, for example, it is noticeable that the ANA logo is much more prominent on the tail fin (compare Figures 11.3b and 11.4), which could aid in raising the awareness of the ANA name. Indeed, developing the findings of Lee *et al.* (2018), it is possible to see how having a more business-oriented tail fin and a body livery that may appeal more to vacation flyers could aid ANA in speaking to both sets of flyers, particularly when compared with the Pokémon planes which probably spoke more to vacation flyers despite the fact that even business people may be spotted playing Pokémon Go.

However, how visible is the Star Wars livery, particularly in countries where observation decks are less common at airports? As noted above, not all passengers are aware that they are travelling on one of the planes that has a special livery. Booking a ticket on a plane with a special livery can also be problematic as the planes are only used on limited routes on particular days, thus further restricting the international exposure, and there is always an advisory that the plane may need to be changed for operational reasons (ANA, 2018). For fans of the popular culture brand, in this case Star Wars, wanting to engage in contents tourism on a plane, there is a possibility of disappointment. Seeing or travelling on the Star Wars planes requires some effort or good luck.

However, I would like to suggest that there is another possible reason for the special popular culture liveries beyond any potential economic benefit to the airline and that this can be found from considering what contents tourism is, who engages with it and how this is done. In the introduction to this book the importance of popular culture is made and that, further to Seaton *et al.* (2017: 4–5) noting popular culture 'by being "of the people," popular culture is not the work of government and there are no distinctions between "art" and "pop culture" or "high" and "low"', emphasizes again the importance of fans and their tourist behaviours. Extending this concept to planes, it would suggest that some fans of the popular culture are choosing to take a particular plane owing to their love of the popular culture rather than a need to fly and visit the destination to which the plane is going. Undoubtedly such people may exist. However, I would like to argue that we should also consider companies as fans. Indeed, rather than asking why a company such as ANA does a special popular culture-related livery, perhaps a more pertinent question would be, why would they not do this?

Whilst we speak of airlines as companies, they, like any company, are made up of people. In a country where contents tourism appears to be so well developed and accepted, even if the term is not yet widely known and there may be different ideas about what it entails, it should come as no surprise that popular culture icons appear on the plane, particularly on the plane of a company which already has a background of using special liveries. In other words, we should be seeing this not as a business strategy but rather an expression of fandom. This is contents tourism where ANA, in the case of the Star Wars liveries, is the fan, rather than a top-down provider of contents tourism as discussed by Butler (Chapter 5) in this book. Just as more conventional contents tourism fans partake in activities and make costumes for other fans to enjoy as well as for their own pleasure, so ANA is creating contents tourism liveries for their own benefit as well as for other fans of Star Wars. Consequently, Figure 11.5 shows that there is an overlap between some popular culture fans, some photographers and some flyers. Within this overlap is the force of contents tourism, which leads to the special liveries being adopted on some planes, and in this case the airline is one of the popular culture fans.

If we accept the view that an airline can be a fan of popular culture, then the planes should not merely be seen as the place of contents tourism but, in the case of some designs, for example the C3-PO plane, as a form as cosplay, as the plane is largely being painted in order to take on the look of the character itself. While Rastati (Chapter 9) discusses cosplay in relation to people within their chapter in this book, I would like to suggest that we need to bring together the concepts of 'wrapping', contents tourism and cosplay to think much more flexibly about who can provide contents tourism and where this can happen.

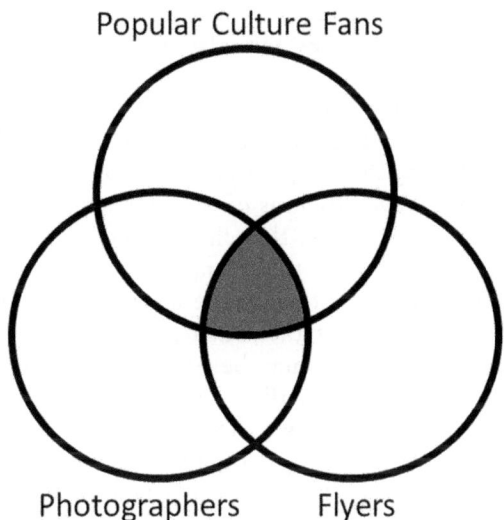

Figure 11.5 The force for contents tourism on planes

Conclusions

The primary aim of this chapter has been to challenge the ways we think about contents tourism, particularly in terms of where it happens, but also in terms of what it is and who does it. As can be seen in the majority of the chapters in this book, contents tourism usually happens at a place after travelling. This further underlines the ideas presented by Seaton *et al.* (2017), where the subtitle of the book it emphasizes that it is about 'Pilgrimages to "sacred sites" of popular culture'. This chapter, however, has looked at contents tourism in relation to the liveries of Japanese planes. Just as contents tourism has been seen to be about activities at a destination, so commercial planes are also primarily used for passengers to get to a destination, rather than for the travel experience itself. However, this chapter has demonstrated that, by using imagery on liveries related to popular culture, Japanese airlines have presented an extension to how we understand contents tourism.

A company such as ANA may get economic benefits from putting contents-related liveries on some of it planes, as its name becomes better known outside of Japan and the company becomes seen as an exciting brand. From a business perspective, the attention given to press releases about these special livery planes may be akin to the impact of having a car in Formula 1 or the Le Mans 24-hour race for car manufacturers; customers may never get to have such exciting cars themselves but using a car that is made by the same company has an associated positive impact. Extending that concept, just as some features of the motor racing cars can be found in a standard road car, so ANA's Star Wars links can be found on its other planes in the form of merchandise for purchase or in the inflight magazine, where a page lists the details (length, top speed, etc.) of ANA's various planes but also includes details about the Millennium Falcon.

However, this chapter has argued that the special liveries in planes are about more than branding and economic benefits. They are about an expression of fandom by the company and the people within it. Although the liveries may be enjoyed by fans of that particular popular cultural brand, for example Star Wars, this chapter has also argued that we should be prepared to consider that companies themselves may be partaking in contents tourism, that whilst the special livery of the plane may seem merely an attempt to attract business or raise awareness of its brand, and this is undeniably may be part of it, it may also be done because it is fun. Accepting this allows us to think more broadly about the design of the plane itself and see that the 'wrapping' of the plane can perhaps best be understood as a form of industrial cosplay. Consequently, this chapter has pointed out that, particularly at or near Japanese airports, it becomes possible to see the airline partaking in contents tourism in plain sight.

Note

(1) Airlines tend to fall into one of three categories; legacy carriers, hybrid carriers and LCC. Most legacy carriers will allow passengers to have a certain amount of baggage allowance and some sort of meal or drink included in the price of the fare. LCCs, on the other hand, tend to require passengers to pay additional charges for any extra services used. Hybrid carriers, as the name implies, fall between the two, often having cheaper services than the legacy carriers, but still offering some included services, such as a baggage allowance, in comparison with LCCs. Based on my experiences of using airlines in Japan, other than a variation in fares, the boundaries between many of the hybrid carriers and the two legacy carriers are particularly blurred.

References

ANA (2015a) Star Wars™ jets (R2-D2™ ANA Jet/Star Wars™ ANA Jet/BB-8™ ANA Jet). See https://www.youtube.com/watch?v=evXqQjpQLjk (accessed September 2015).
ANA (2015b) Fans with the R2-D2™ ANA Jet. See https://www.youtube.com/watch?v=BIHHFjqw1J8 (accessed December 2015).
ANA (2016) Star Wars: Main title arranged by ANA for Star Wars Day. See https://www.youtube.com/watch?v=D2YO3VkRncc (accessed May 2016).
ANA (2017) The C3-PO ANA Jet will fly services under pre-set summer schedule from July 18, 2017. See https://www.ana-sw.com/en/news/245 (accessed June 2017).
ANA (2018) Star Wars Jets. See https://www.ana-sw.com/en/jets (accessed May 2018).
Budd, L.C.S. (2012) The influence of business models and carrier nationality on airline liveries: An analysis of 637 airlines. *Journal of Air Transport Management* 23, 63–68.
CAPA (2018) ANA & JAL dominate Japan's domestic airline market: Record traffic. See https://centreforaviation.com/analysis/reports/ana--jal-dominate-japans-domestic-airline-market-record-traffic-413333 (accessed May 2018).
Craig, T. (2017) *Cool Japan: Case Studies from Japan's Cultural and Creative Industries*. Ashiya: BlueSky.
Doi, T. (1989) *The Anatomy of Self: The Individual versus Society* (trans. M. Harbinson). Tokyo: Kodansha International.
Hendry, J. (1993) *Wrapping Culture: Politeness, Presentation and Power in Japan*. Oxford: Clarendon Press.
Hood, C.P. (2006) *Shinkansen: From Bullet Train to Symbol of Modern Japan*. Abingdon: Routledge.
Kageyama, Y. (2006) Cuteness a hot-selling commodity in Japan. See http://www.washingtonpost.com/wp-dyn/content/article/2006/06/14/AR2006061401122.html (accessed June 2006).
Kinsella, S. (1995) Cuties in Japan. In L. Skov and B. Moeran (eds) *Women, Media and Consumption in Japan* (pp. 220–254). London: Curzon Press.
Lee, J., Yi, J., Kang, D. and Chu, W. (2018) The effect of travel purpose and self-image congruency on preference toward airline livery design and perceived service quality. *Asia Pacific Journal of Tourism Research* 23 (6), 532–558.
MLIT (2016) *Kōkū Ichiran*. Ministry of Land, Infrastructure, Transport and Tourism. See http://www.mlit.go.jp/koku/15_bf_000310.html (accessed June 2016).
Raguraman, K. (1997) Airlines as instruments for nation building and national identity: Case study of Malaysia and Singapore. *Journal of Transport Geography* 5 (4), 239–256.
Seaton, P., Yamamura, T., Sugawa-Shimada, A. and Jang, K. (2017) *Contents Tourism in Japan: Pilgrimages to 'Sacred Sites' of Popular Culture*. New York: Cambria Press.

Thurlow, C. and Aiello, G. (2007) National pride, global capital: A social semiotic analysis of transnational visual branding in the airline industry. *Visual Communication* 6, 305–344.
Wang, S.W. (2015) The experience of flying with Hello Kitty livery featured theme jet: Moderating effects of destination image. *Current Issues in Tourism* 18 (2), 99–109.
Wang, S.W. and Ngamsiriudom, W. (2015) Celebrity livery featured aircraft, the Maneki Neko (fortune cat) of airlines. *Journal of Air Transport Management* 42, 110–117.

12 Breaking Benjamin: A Woman's Pilgrimage to New Mexico

Stefanie Benjamin

I identify as a popular culture fanatic with a passion for television and movies. Growing up as a child, my family and I shared many fond experiences around the television, quoting and re-enacting scenes from various television shows and films. When I decided to pursue a career as an academic, this passion spilled over into my professional life. The topic of my Master's thesis was *The Andy Griffith Show* and the Mayberry Days Festival in Mount Airy, North Carolina. This festival inspired my research interest in film-induced and contents tourism. I had never been a fan of the show, nor was I interested in the economic impact of the series on Mount Airy. Instead, I was interested in the potential sociocultural effects that movies, television shows and books can have on a place, destination and culture. During my PhD studies, I developed a strong desire to investigate the effects of a show in which I was personally invested. During my first year as an assistant professor, just such an opportunity presented itself. Adopting an autoethnography approach, I explored the tourism industry phenomenon in Albuquerque, New Mexico inspired by the television show *Breaking Bad*.

Originally aired on the AMC network for five seasons, from 20 January 2008 to 29 September 2013, *Breaking Bad* follows the life of a fictional character, Walter White, a high school chemistry teacher who becomes a drug dealer after being diagnosed with terminal cancer. Filmed and set in Albuquerque, the city continues to promote and market itself as a tourist attraction through its Convention & Visitors Bureau with several *Breaking Bad* tours, including filming locations for the prequel, *Better Call Saul*. Ten years after the premiere of *Breaking Bad*, Albuquerque continues to attract significant revenue from tourists dedicated to 'never letting Walter White's image fade' (Vanderhoof, 2018).

My tourist pilgrimage began in June 2016 with two tours in Albuquerque: a *Breaking Bad* Biking Tour and a Recreational Vehicle (RV) Tour. During the tours, I adopted an autoethnography approach to explore how emotional and sensory contents are connected. I used a social media application, Snapchat, to document my lived experiences with the tours. My posts were compiled into a 'story' showcasing my personal video diaries and snap chats. The footage exemplified my lived experiences of being a female film tourist and addressed my personal and emotional relationship with film tourism from a sensory/embodied ethnography framework (Pink, 2015).

In this chapter, I use a visual autoethnography methodology and a storytelling approach and data from social media platforms, field notes and journals to share my experiences, motivations and travel behaviours as a female film-fanatic tourist. Goodall (2000) advocated for a more feminine communication style in academia and emphasized rapport building through listening and observing (e.g. by engaging in personal reflection about meanings) rather than problem solving. Accordingly, I embarked on a media pilgrimage (Norris, 2013) to examine the intimacy of travelling as a solo female film-induced tourist and the intersectionality of contents tourism by mimicking imaginary hedonistic characters and exploring landscapes associated with the American television show *Breaking Bad* during the summer of 2016.

Commodifying *Breaking Bad* in Albuquerque

The most powerful visual medium for creating and shaping images today is video – specifically, films and television shows (Roesch, 2009). Television is commonly regarded as the most powerful advertising medium because consumers can view products in realistic scenarios. Television series and films portray products, experiences and locations in ways that motivate tourists to seek them out. Location attributes are realistically transmitted through televised images, which results in a globally accepted sense of portrayed places (Massey, 1994).

Television shows and films have both positive *and* negative effects on the communities portrayed in them. Riley (1998) argued that film-related tourism creates numerous benefits, including the creation of organized tours, memorabilia sales, new uses for community sites, expansion of community festivals and exposure for establishments used as film locations. In the case of *Breaking Bad* tourism in Albuquerque, several tours of film locations and souvenir shops have been established (see Figure 12.1). Tourists are flocking to locations that, before the show aired, were relatively unknown. For instance, the Albuquerque Convention & Visitors Bureau created a website to help tourists navigate between physical filming locations and notable sites from the show via a personal walking tour (https://www.visitalbuquerque.org/about-abq/film-tourism/breaking-bad/). Several paid tours, ranging from US$60 to 75, also include an Albuquerque Trolley Tour, *Breaking Bad* RV Tour, Biking Bad Tour, and limousine tour with the Candy Lady, who produced and provided the fake 'blue meth' for the show.

Figure 12.1 *Breaking Bad* 'Heisenberg/Goodman 2016' t-shirt in Old Town. Many shops within Old Town featured *Breaking Bad* and *Better Call Saul* souvenirs ranging from shot glasses to underwear. Author's photo

One major negative effect of the show is Albuquerque's association with drug use. Instead of focusing on the real drug problem that Albuquerque faces with opiates, Vince Gilligan, the creator, director and writer of *Breaking Bad*, substituted methamphetamine as the drug of choice (Vanderhoof, 2018). Sandy,[1] a sales clerk in a t-shirt store located in the tourist centre of Old Town, explained:

> Being a local we are touchy about it ... I was born and raised here and I'm very proud of this place [...] we aren't perfect though. We are close to the border and we always had problems with drugs here because they are transported to us. We have two main highways that intersect in the middle of the city [...] a lot of things are transported through here [...] when I talk to people globally, some of them are seeing a problem with meth. (Personal communication, 2016)

Sandy continued:

> People either hate this show or they love it [...] But when people say it is a meth show and don't want anything to do with it, then I say, 'You are cutting yourself short' [...] then I say, 'You can watch *Better Call Saul*'.

Interestingly, even though Sandy expressed concerns about Albuquerque's association with drug use thanks to *Breaking Bad*, she was quick to defend the show and promote the prequel, *Better Call Saul*, which also revolves around drug production and unethical behaviour.

Figure 12.2 Card from a fan of the 'Blue Sky' *Breaking Bad* doughnut hanging in the Rebel Donut shop in Albuquerque. Author's photo

This association with drug use has yet to affect the number of tourists making the pilgrimage to Albuquerque; in fact, it can be argued that it has increased tourists' fascination with the city. Although *Breaking Bad* ended in 2013, tourists from across the nation and around the world come to Albuquerque to taste 'blue meth' doughnuts and re-enact scenes from the show (Vanderhoof, 2018; see also Figure 12.2).

The ability to stream all seasons of *Breaking Bad* on platforms such as Netflix has increased the show's popularity. Sandy acknowledged that '*Breaking Bad* put Albuquerque on the map globally'; she said that numerous tourists from all over the world, including Norway, Netherlands, Germany, France, Italy, Hungry, Poland and Spain, have visited her store thanks to 'streaming platforms' (personal communication, 2016). Adopting an academic and touristic lens, I decided to explore how Albuquerque is sustaining the *Breaking Bad* tourism phenomenon via film-induced and contents tourism.

Conceptual Design: Contents Tourism and Sensory Autoethnography

Film and contents tourism

Television shows and films enable people to be transported to other times and places. Often, tourists seek out environments and experiences that mirror what they have seen on the silver screen or their television sets. For instance, in *The Tourist Gaze*, Urry (1990: 3) suggested that:

> places are chosen to be gazed upon because there is an anticipation, especially through daydreaming and fantasy, or intense pleasures, either on a different scale or involving a different sense from those customarily encountered. Such anticipation is constructed and sustained through a variety of non-tourist practices such as film, TV, literature, magazines, records, and videos which construct and reinforce the gaze.

The tourist gaze can be directed to certain features that are unique and therefore distinguish the 'site/sight' of the gaze from others. The properties of a movie location, whether scenic, historical or literary, qualify as potential icons for tourists (Riley, 1998). However, these icons are hard to measure in terms of whether a tourist is visiting the location based on their fascination with the movie or a willingness to venture to the film site when visiting the destination for other reasons.

This is known as film-induced tourism, which Beeton (2005: 9) defined as 'on-location tourism that follows the success of a movie made (or set) in a particular region'; the concept 'incorporates aspects of disciplines such as sociology and psychology, as well as industry-based sectors from film making through to destination marketing, community development and strategic planning'. Although film-induced tourism is somewhat limited, television series are filmed and screened over longer periods of time, with possible syndication opportunities (Beeton, 2005). This long-term exposure enables viewers to develop stronger relationships with the stories, characters and settings, thereby reinforcing their desire to visit the locations/regions where series are filmed and to see television crews and actors on-location.

Expanding on film-induced tourism, contents tourism (*kontentsu tsūrizumu*) is defined as 'travel behavior motivated fully or partially by narratives, characters, locations, and other creative elements of popular culture forms, including film, television dramas, manga, anime, novels, and computer games' (Seaton *et al.*, 2017: 3). Within this narrative world, however, Takayoshi Yamamura has posited that there are three major actors in contents tourism – creators, fans, and the local community – which, collaboratively, help to establish spaces and places that encourage touristic pilgrimages that elicit emotive bonds (Yamamura, 2015: 75–80). Additionally, contents tourism encourages tourists to 'access and embody "narrative worlds" that are evolving through "contentsization", namely the continual process of the development and expansion of the "narrative world" through both mediatized adaptation and tourism practice' (see Yamamura's Introduction to this volume).

An example of contents tourism is seen with the novel and film, *Wild*. In 2012, author Cheryl Strayed wrote a biography of her struggles with divorce, drug addiction and loss which motivated her to hike, as an un-trained, novice hiker, along the Pacific Crest Trail (PCT). This *New York Times* bestseller was adapted into an Oscar-nominated film *Wild* staring Reese Witherspoon in 2014. These two mediums influenced thousands of novices, mostly female hikers to embark on the same journey as Strayed, causing 'The Wild Effect' (Trageser, 2015). This effect had negative environmental and life-threating issues since tourists were wanting to re-enact this pilgrimage and embody similar transformational experiences to Strayed. Consequently, Strayed partnered with several associations and organizations to caution hikers and listed safe and responsible strategies for hiking the PCT. This 'Wild Effect' was parodied in the reboot of *Gilmore Girls* showing lead character, Lorelai Gilmore, setting out to embark on the same hike after reading *Wild*. The emotive components from both literary and film spheres influenced Lorelai to try and hike the PCT on her own pilgrimage, hoping to share similar life-changing experiences. However, spoiler alert, her journey never transpired; unlike Strayed, who hiked nearly half of the 2650-mile trail, Lorelai never made it past the park ranger. However, they both shared an important commonality: it was more about their emotional journey than the physical one.

Contents tourism goes beyond a 'single notion of film' and intersects with the complexity of human nature and tourism (Beeton, 2016: 31). Furthermore, it transcends fans' shared memories and is a series of touristic experiences motivated by contents. Tourists are (re)living, (re)embodying and (re)experiencing emotive sensations. This (re)contentsization of the narrative world allows for touristic lived experiences that embrace sensory components consumed by making the pilgrimage to such narrative landscapes. My personal pilgrimage to the narrative world of *Breaking Bad* was informed by the lens of sensory autoethnography, where unlike Strayed or Lorelei, I was interested in escaping into the world of *Breaking Bad*.

Sensory autoethnography

According to Ellis *et al.* (2011: 274), 'autoethnography is one of the approaches that acknowledges and accommodates subjectivity, emotionality, and the researcher's influence on research, rather than hiding from these matters or assuming they don't exist'. Accepting reality as multifaceted, autoethnography questions the dominant scientific paradigm, thereby allowing a sharing of unique, subjective and evocative stories that contribute to understandings of the social world (Wall, 2006). The researcher's experiences are highlighted by revealing details of the self while interacting with the phenomenon in context. The use of thick descriptions to describe these experiences is essential in illuminating a deeper understanding of the phenomenon (Bochner & Ellis, 2002). As Schaeperkoetter (2017: 132) posited, 'autoethnography allows the researcher to reflexively examine, through self-observation, the social forces shaping their experiences'. Thus, ethnographic experiences are 'embodied' as researchers negotiate the spatial context of the field (Pink, 2015).

An interdisciplinary focus on the senses has emerged during the early 21st century. For instance, Pink (2015) argued for traditional ethnographic methodologies to be reinterpreted through attention to sensory experience, including embodiment. Embodied ethnography enables researchers to interpret the world through their own positioning, visibility and performance while engaging in participant observation (Coffey, 1999). The 'experiencing, knowing, and emplaced body' is central to sensory ethnography whereby researchers reflect on bodily sensations 'to conceptualize their meanings and intellectual meanings' (Pink, 2015: 21). With this in mind, I used sensory ethnography to explain my lived experiences as a film-tourist and how I embodied tourism through two select *Breaking Bad* tours. I kept several field journals (both visual and written), took copious amounts of photographs, spoke informally with stakeholders, including store managers, residents and tourists, and also used the Snapchat social media platform to document my time spent in Albuquerque (Figure 12.3). Through triangulation among these data sources (Saldaña, 2015), I was able to create a narrative of my pilgrimage using two lenses eliciting emotive bonds and access to (re)contentsization of the narrative world of *Breaking Bad*.

My Pilgrimage to Albuquerque

I identify as an able-bodied, White, heterosexual, cisgender woman who had the flexibility to embark on a pilgrimage as part of my research program. At times, it was difficult to separate the researcher and tourist

Figure 12.3 Part of my Snapchat story where I was 'waiting for Jesse to return'. Author's photo

lenses. Thus, this chapter is a blended account where I try to explain my observations and feelings of wearing two lenses simultaneously. Furthermore, I acknowledge that not everyone would be able to physically participate in the Biking Bad tour, or be able to financially afford to take this pilgrimage. Having made these privileges fully transparent, I share an account of my pilgrimage and associated lived experiences during my trip to Albuquerque in the summer of 2016.

Biking Bad tour

Routes Bicycle advertises its Biking Bad tour as a 'slow travel' experience of the locations, sights and sounds of *Breaking Bad* in addition to the 'vivid southwestern landscapes of the city and show' (Routes, n.d.). It is marketed as a form of 'intimate' and 'interactive' engagement with locations featured in the show with the help of professional tour guides on 'five distinct tour routes, each with a unique perspective and sequence of locations to be explored' (Routes, n.d.). This tour enables tourists to 'experience the *bad* side of Albuquerque ... by bike' (Routes, n.d.) on a seven-mile guided tour of 11–13 major locations in and around the urban core of Albuquerque. I participated as the only guest on this tour on 21 June 2016 from 10:00 am to 1:00 pm I paid $65 (US) and was shown clips from *Breaking Bad* before we ventured to each site location to help me 'get into the mood'.

Once I was fitted for my bicycle and provided with a helmet, we were ready to leave Old Town. My tour guide, a young woman named 'Monica' who had recently moved to Albuquerque, was not only incredibly knowledgeable about *Breaking Bad*, but was a huge fan as well (Figure 12.4). While riding behind her, I noticed that there were no advertisements or marketing materials for Routes Bicycle anywhere on our bicycles or helmets. It was as if I was venturing into town with a friend; we discussed the plots and characters of our favourite television shows and enjoyed bouncing theories off each other.

One of my favourite locations that we visited was Jesse's aunt's house, which eventually became Jesse's house in the show. It is located in an affluent neighbourhood surrounded by large, well-kept homes and perfectly manicured lawns. To my surprise, the house of Chuck McGill, a character from the prequel *Better Call Saul*, was around the corner. Monica explained that it was Vince Gilligan's vision to maintain authenticity and to ensure that the sites/locations were as 'real' as possible. Thus, Jesse's house was within biking distance of the real Dog House, where Jesse sold meth and a real place to score cheap and greasy food. Gilligan also included the Dog House in *Better Call Saul*, along with many other filming locations familiar to *Breaking Bad* fans.

While at Jesse's house, we noticed several people stopping their vehicles to take photographs or posing in front of the home. They conversed

Figure 12.4 My Routes Bicycle 'Biking Bad' tour guide (pictured left) and me before our seven-mile biking tour. Author's photo

with me as if I was a local, asking questions about other film location sites. It was interesting to hear their reasons for stopping in front of the home: some were huge fans making the effort to visit locations on their road trip to California, whereas others were just seeing what the 'fuss was all about'. As for me, it felt surreal to be physically standing in front of a home that I felt oddly connected with. Although Monica was also relaying the history behind the neighbourhood and the architecture of the homes, I was more interested in the potential hypothesis of Jesse living so close to Chuck McGill and a crossover of their two worlds. Was this intentional on Gilligan's part, since he had insisted on using real street names and businesses to maintain authenticity? These thoughts raced through my head as we headed to the Dog House, then downtown toward Tuco's headquarters, which is a real coffee shop named Java Joe's.

On our bicycle ride over to the downtown Albuquerque locations, Monica shared a conversation she had had with a German tourist during a previous tour. She said that he felt like *Breaking Bad* was a commentary on the US healthcare system and that 'this sort of thing wouldn't be an issue with places like Germany that have universal healthcare for their residents'. It was a perspective that I had never considered. Later, I discovered that Bryan Cranston, the actor who played *Breaking Bad*'s main character, Walter White, commented during a 2011 *Rolling Stone* interview, 'If we did have universal health care five years ago, the show might not have worked … Thank God Obamacare wasn't in play five years ago. Whew!' (Leonard, 2011: para. 4).

During the tour, I experienced somewhat of a personal identity crisis that other academic researchers may also experience during data collection. As a tourism studies scholar, it was difficult to put down my academic lens and allow myself to simply be in the moment and enjoy being a tourist. Instead, my brain was in overdrive thinking about social equity issues and how a television show could serve as a powerful platform to advocate for change. I then began chatting with Monica about how the Walter White character, as a US public school teacher, had to work a second job at the A1A Car Wash just to make a decent living wage. When he could not afford the costs associated with his terminal cancer diagnosis, he resorted to cooking meth to support his family. I had a similar conversation around inequity issues with Sandy. She said:

> It shows what a typical person could become if you are out in [between] a rock and a hard spot. It says a lot about teachers, because teachers may have to go to extremes because [they are] not making the right amount of money … and they could be put in that position.

Sharing feelings about real-life issues portrayed on the show helped to create an experience that felt more intimate than a typical film location tour. This was an unexpected aspect of the tour that may be a general attribute of smaller-sized approaches to tourism. My academic and touristic lenses merged together and were welcomed into this space. Perhaps also, the senses generated from physically riding a bicycle, feeling and smelling the air, and having the freedom to stop and start the tour when I wanted to, all combined into a truly unique tourism activity. I was given opportunities not only to pose for photos at key locations and reenact scenes (Figure 12.5), but also to discuss important political issues that related to plot lines in *Breaking Bad*. Our conversation led us to one of our last locations: a parking garage where Walter White tried to assassinate another character, Gus. This led to another important discussion about gun access and control in the USA while I posed as Walter White and reenacted the 'It's done' scene.

I enjoyed engaging in deep discussions around gun control, healthcare and salary inequities with Monica while also being able to embody and reenact scenes organically as I visited the film locations. As Riley (1998) stated, tourists and residents both feel a sense of ownership with regard to film locations. In my field notes, I documented how 'reliving and recreating the scenes made me feel like I owned part of the show' and perhaps motivated my strong desire to discuss the sociocultural impacts of *Breaking Bad* with other fans and tourists. Until that point, I had not realized the show's role in generating deeper dialogue around various political issues in the USA. This sentiment led me into an academic haze and generated many unanticipated feelings eliciting powerful emotions.

Figure 12.5 Parking garage reenactment on Walter White's cell phone. Author's photo

Breaking Bad RV tour

The *Breaking Bad* RV tour was a three-hour 'interactive' tour that transported us in a replica of the RV in which Walt and Jesse cooked meth on the show (Figure 12.6). The tour enabled tourists to experience the sights and sounds of the show with stops at 20 locations, including lunch at the fictional Pollos Hermanos, a real fast-food restaurant called Twisters. On the tour, guests had opportunities to 'interact with the hosts' who had 'worked on the show' and to win prizes for answering trivia questions correctly. My tour began at 9:00 a.m. on 23 June 2016; I paid $75 (US), and was joined by 12 tourists visiting from Germany, England, Australia, Albuquerque, Indiana and Texas.

Tour participants met in Old Town an hour early owing to extremely hot temperatures and a lack of good air conditioning in the RV. As we were waiting for our tour guide to arrive, my group was pretty distant; no one really made an effort to talk. Several thoughts ran through my head: Am I in the right spot? Will this be worth it? Once the RV pulled into the parking lot, we all perked up and were ready for the adventure. Our tour guide was a man in his forties and the owner of the operation. He later shared with me that he had started the tour as kind of a joke between friends and had not anticipated that *Breaking Bad* would turn Albuquerque into a film tourism destination. It did take him longer than expected to

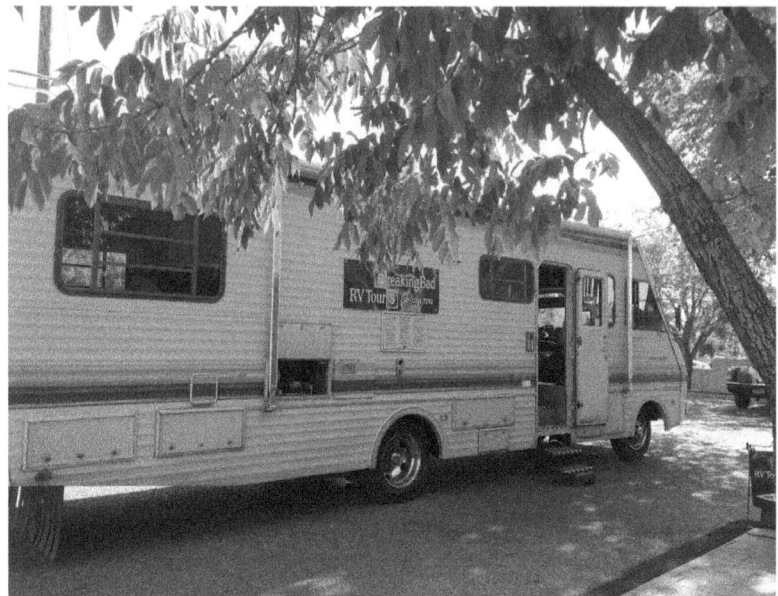

Figure 12.6 *Breaking Bad* RV Tour located in Old City in Albuquerque. Author's photo

obtain the copyright for the name *Breaking Bad* RV Tour, but he felt that it had been worth the investment.

The RV included a replica of a meth cooking station, photos from actors in the series, and various artefacts related to the show, including a dried up lily of the valley plant and a pink teddy bear. The van was quite warm, even though it had been advertised as having air conditioning, and I was only semi-comfortable with my seating arrangement, squeezed next to a couple in their fifties from Australia. Our tour guide, using a microphone, explained which locations we were visiting and why we would not be able to physically stop owing to traffic or private property restrictions. We began by visiting Jesse's house and some downtown locations similar to the Biking Bad tour; however, when we started venturing out toward Northeast Heights, our adventure became truly different.

In Northeast Heights, we visited three key locations in *Breaking Bad*: Walter White's house, Los Pollos Hermanos and the A1A Car Wash. Although it was exciting to see Walter White's house in person, it was truly a disappointment. Our tour guide claimed to be 'friends' with the owner of the home, who allowed us to take photos, as long as we stayed behind our parked RV on the road. She had dealt with major vandalism-related issues, including tourists reenacting throwing a pizza onto her roof, stealing rocks from her yard and selling them on eBay and swimming in her pool. Consequently, she was diligent about controlling how people visited her home; some years after our tour, she built a fence.

After posing in front of the home and taking several photos, we boarded the RV and stopped off at Los Pollos Hermanos (a.k.a. Twisters) for lunch. Twisters had signage for Los Pollos Hermanos as depicted on the show to help tourists and visitors recognize that this was the location owned by the character Gus. I reenacted various scenes while eating my lunch but did not see any fellow RV tourists embodying characters (Figure 12.7).

At no point on the tour did I feel like I had when I was on the Biking Bad tour. The entire time, I felt like I was part of some sleazy, low-budget tour; instead of blending in as a local, I was standing out like a cheesy tourist. My feelings grew even stronger at our next stop: the A1A Car Wash where Walter White's character worked, and which he later bought as a money laundering operation. The real operation, Octopus Car Wash, had signed *Breaking Bad* posters lining the hallway and cashiers who were instructed to say, 'Have an A-1-A day', as quoted from the show. After we walked through the site, we gathered around the RV to watch a 'blue meth' demonstration where the tour guides lit greenish smoke out of the RV to resemble a 'cooking session' (Figure 12.8). This incredibly embarrassing and weak attempt at entertainment was followed by an awkward, staged group photograph to be used as a marketing tool for the tour company's social media sites.

Figure 12.7 Posing as Heisenberg at Los Pollos Hermanos (known as Twisters in real life). Author's photo

Figure 12.8 Tour member taking photos of the 'meth cooking session' from our RV at the A1A Car Wash (known as the Octopus Car Wash in real life). Author's photo

Overall, the tour left me feeling like it had been created by Saul Goodman, the fictional scam artist turned scam lawyer from *Better Call Saul*. The tour operator made no attempt to establish a sense of authenticity or intimacy; instead, it felt more like a way to capitalize on and commodify tourists' enthusiasm and love for the show. Perhaps my dual tourist and academic lenses influenced my feelings about this tour, but I did not appreciate feeling like someone was taking advantage of my love for the show to make a buck. I recognize that there were more people on this tour than on the Biking Bad tour and thus the experience would, by definition, be less intimate; yet our tour guide made no attempt to get to know us or build any type of community amongst us. I felt like the entire operation was a disingenuous attempt to capitalize on a mass tourism opportunity instead of a sincere attempt to create a film-induced experience.

Better Call Saul: Continuing the Pilgrimage?

My experiences reveal that it is essential to promote and implement film-induced or contents tourism experiences in an intimate and authentic manner. Several residents and various stakeholders expressed similar sentiments during discussions throughout my pilgrimage. Fans feel a strong desire to ensure that Walter White's image lives on. Tours and tourism-related activities such as Biking Bad help fans fulfill this desire. Yet how sustainable is *Breaking Bad* tourism to Albuquerque? In the contents tourism industry, incorporating spin-offs to help sustain the popularity of a film series and establishing links between people and physical film location sites are crucial to sustainability (Yamamura, 2015). Susan and I chatted about the sustainability of *Breaking Bad* in terms of Albuquerque tourism. She shared:

> Will *Breaking Bad* die eventually? I think so ... but with Saul ... a tourist shared [an idea] with me: wouldn't it be cool if at one of the restaurants on Central Avenue that was featured in *Breaking Bad* and *Better Call Saul*, if Saul ends up sitting in that restaurant in a booth, and the guys from *Breaking Bad* are sitting next to him? Really, that is what Albuquerque wants, too. For the shows to combine and leave the tourist wanting more.

Perhaps the tourist that Susan quoted was correct, and a crossover is possible. Regardless, I hope that future tours created to combine these two worlds are managed similarly to Biking Bad. Maintaining small tour sizes ensures deeper dialogue, not only about the show or characters, but also about socio-cultural issues beyond the architectural styles of the homes and neighbourhoods. This could be a unique attribute that enables tourists to have meaningful experiences. Adopting such an approach to contents tourism can potentially reveal how films and television shows can serve as powerful platforms for advocacy around social issues. Like the German tourist who drew on the plot of *Breaking Bad* to discuss larger social issues related to the US healthcare system, participants have the responsibility and power to shape discussions during film-induced tourism experiences. Hopefully, more tourists will begin to feel comfortable sharing their stances on socio-cultural components of shows and use tours as opportunities to engage in difficult, but necessary dialogue about important issues, including the reality of drug use, health care, wages and gun laws.

Note

(1) All names of local informants have been anonymized.

References

Beeton, S. (2005) *Film-Induced Tourism*. Clevedon: Channel View Publications.
Beeton, S. (2016) *Film-Induced Tourism* (2nd edn). Bristol: Channel View Publications.
Bochner, A.P. and Ellis, C. (2002) *Ethnographically Speaking: Autoethnography, Literature, and Aesthetics*. Walnut Creek, CA: Alta Mira Press.
Coffey, A. (1999) *The Ethnographic Self: Fieldwork and the Representation of Identity*. London: Sage.
Ellis, C., Adams, T.E. and Bochner, A.P. (2011) Autoethnography: An overview. *Historical Social Research/Historische Sozialforschung* 36 (4), 273–290.
Goodall Jr, H.L. (2000) *Writing the New Ethnography* (Vol. 7). Lanham, MD: Alta Mira Press.

Leonard, A. (2011) Bryan Cranston on the joy of cooking meth, Obamacare and *Malcolm in the Middle*. *Rolling Stone*. See https://www.rollingstone.com/culture/culture-news/bryan-cranston-on-the-joy-of-cooking-meth-obamacare-and-malcolm-in-the-middle-199972/ (accessed January 2019).

Massey, D. (1994) *Space, Place, and Gender*. Cambridge: Cambridge University Press.

Norris, C.J. (2013) A Japanese media pilgrimage to a Tasmanian bakery. *Transformative Works and Cultures 14*, 1–16.

Pink, S. (2015) *Doing Sensory Ethnography* (2nd edn). Los Angeles, CA: Sage.

Riley, R. (1998) Movie induced tourism. *Annals of Tourism Research* V (I), 919–935.

Roesch, S. (2009) *The Experiences of Film Location Tourists*. Bristol: Channel View Publications.

Routes (n.d.) Biking Bad Tour ABQ. See https://routesrentals.com/blogroll/tours/specialty-bike-tours-abq/biking-bad-tour/ (accessed November 2018).

Saldaña, J. (2015) *The Coding Manual for Qualitative Researchers*. Los Angeles, CA: Sage.

Schaeperkoetter, C.C. (2017) Basketball officiating as a gendered arena: An autoethnography. *Sport Management Review* 20(I), 128–141.

Seaton, P., Yamamura, T., Sugawa-Shimada, A. and Jang, K. (2017) *Contents Tourism in Japan: Pilgrimages to 'Sacred Sites' of Popular Culture*. Amherst, NY: Cambria Press.

Trageser, C. (2015) The Wild effect: Hikers are flooding the Pacific Crest Trail. See https://mashable.com/2015/05/17/pacific-crest-trail-wild-effect/#ltKXMd9DpOqs (accessed January 2019).

Urry, J. (1990) *The Tourist Gaze: Leisure and Travel in Contemporary Societies*. London: Sage.

Vanderhoof, E. (2018) Ten years later Albuquerque is still *Breaking Bad's* town. *Vanity Fair*. See https://www.vanityfair.com/hollywood/2018/01/albuquerque-breaking-bad-tourism-10th-anniversary (accessed January 2019).

Wall, S. (2006) An autoethnography on learning about autoethnography. *International Journal of Qualitative Methods* 5 (2), 146–180.

Yamamura, T. (2015) Contents tourism and local community response: *Lucky Star* and collaborative anime-induced tourism in Washimiya. *Japan Forum* 27 (1), 59–81.

13 From Banjo to Basho: Poets, Contents and Tourism

Sue Beeton

Introduction

Poetry emotionally connects us not only to our interior world, but also to other people and even places. As such it can be a deeply powerful element of the tourism experience. While modern poets may abound in today's pop culture in various forms from the beat poets of the 1950s to the hip-hop and slam poetry of this century, it seems to be the more traditional poets who engage our collective soul, which may simply be a reflection of the time it takes for this to occur along with the role that memory and nostalgia play, or something deeper and more elemental. As an Australian, the poet A.B. 'Banjo' Paterson (1864–1941) presented me with many images and stories that remain central to my Australian-ness and are often referred to in our tourism experiences. Iconic poems such as *The Man from Snowy River* and *Waltzing Matilda* are not only situated in the Australian bush but also in the psyche.

Many Australians, including migrants, still connect with these poems, along with countless visitors who see Australia as a final frontier where such adventure is still possible, giving them an insight into what is arguably an imagined Australia, yet one that provides them with a deeper touristic experience through the emotions, romance and adventure depicted in our epic poetry (Beeton, 2016).

In a not dissimilar way, the Japanese poet, Basho Matsuo (1644–1694) takes us through Japan via his famous poetic travelogues, the most renowned being *Oku-no-Hosomichi* (*Narrow Road to the Interior*), considered to be one of the great works of Japanese literature. In this chapter, we explore the personal and emotional connections of poetry via two seemingly unconnected poetic forms, and discuss how this contributes to the concept of 'contents tourism'.

Poetry as Contents Tourism

While the term itself was developed in Japan, *contents tourism* not only applies to that country or culture, and can be witnessed around the world (Beeton *et al.*, 2013). In fact, there are many international scholars studying the deep connections we have with 'content' from Reijnders' *Locating Imagination* research in the UK and Europe to my own consideration of the emotional role of the moving image (Beeton, 2015; Reijnders, 2011). In the chapter 'The mediatisation of culture: Japanese contents tourism and pop culture', Beeton *et al.* (2013) discuss the defining feature of the narrative quality of contents tourism, which is further developed in this publication by many of the authors. As Seaton (Chapter 1) notes in this book, contents tourism comprises three basic elements: narrative, character and multiple media. Certainly, poetry provides us with many deep and varied narrative qualities, and the two cases I look at here have been presented via numerous media, from oral presentations to the printed word, film and via social media, as well as being presented at sites, monuments and events.

Many aspects of popular culture come together to create contents tourism, from film and TV to music and art, yet poetry has not been adequately explored, in spite of it being a creative form that engages us on such a deeply emotional level that it can readily create highly emotional tourist experiences. While poetry remains part of the popular narrative today, the distance of time has created works of great emotional resonance; indeed, they take on a national resonance.

Poetry-induced Tourism

Often seen as a subset of literary tourism, poetry is a highly imaginative and emotive form of expression and entertainment, providing the fan with what is often an emotional experience and connection, not only to the poet but also to the material. However, not all poetry forms are the same, varying from the long epic poems full of detail, description and narrative to the tiny, delicate haiku, where a few words infer worlds, both inner and outer. While both of these creative forms remain relevant, despite their age, Australian 'bush poetry', which is often expressed in epic terms, is often denigrated, yet the narratives are highly popular, informing a national (predominantly white) Australian consciousness. The Japanese haiku also informs that country's national consciousness, hence the power they both can have for visitors to these respective countries. Just like participating in a culture via learning the language, cooking and eating local food and listening to local music, reading and following the stories in the literature of a place gives tourists a connection that simple passive gazing cannot.

In 1982 I participated in my first horseback safari which was based on a movie inspired by one of Australia's most iconic poems, *The Man from Snowy River*, one that is still recited around the campfires of the Australian bush today. The power behind experiencing some of the living elements of this poem via tourism was, for me, a numinous experience changing the direction of my life. In 1989 I began working on these tours as a guide, ultimately moving into tourism academia where I was able to continue exploring the power of popular culture on both a practical and a theoretical level.

Such studies led me to other cultures, including Japan, and to the various artists, film makers and poets who define so much of that country's background. The ways in which places are defined through art and artistry in Japan are endless, and the landscape itself inspires poetic expression.

As a teenager keen on expressing myself through poetry, I recall being introduced to the subtle simplicity of the Japanese haiku, at a time when I was particularly open to hearing the emotional depth that only poetry can bring. I recently had the joy of following the trail of Basho Matsuo's *haibun* (short prose passages with haiku) that touch me deeply, but I suggest quite differently to my Japanese friends, primarily as I am experiencing this as an outsider, a visitor, a tourist.

While the epic poetry of A.B. 'Banjo' Patterson is very different from Basho Matsuo's delicate haiku, and separated in time, distance and form, they both resonate within me – one tells of my own physical life, the other my inward journey and sense of Japan. Simply put, I am considering the poetry of my homeland and how it affects my domestic travel experience alongside the poetry of another place and culture and how it informs that international travel experience.

As a child, listening to poems from Banjo Patterson such as *The Man from Snowy River* moved me significantly, and when I had the chance to experience the country depicted in the poem in the actual manner of the poem (via horseback), my emotional response was powerful. In the same way, reflecting on Basho's work as I travelled Japan in his footsteps, I felt a connection to place that I had not hitherto experienced in Japan.

This chapter is a reflection on this powerful connection between poetry and place, using myself as the 'data' via autoethnography, viewed through a contents tourism theoretical lens. In fact, by looking at poetry in this way, we are free to move away from the 'literary' silo, opening poetry up to travel and tourism.

Autoethnography

Autoethnography (see also Chapter 12 by Benjamin) is a research method and approach that is not well understood, and at times not handled appropriately even by its proponents. This approach is often taken in cases where it is not possible to access deeply personal information, such as in cases of abuse and violence (Beeton, 2008, 2016). By turning the ethnographic gaze inwards, we can then understand the larger experiential world (Denzin, 1997: 227). However, when we use ourselves as the data, deep reflection and exposure are required, otherwise the work can fall into the realms of a simple memoir.

While tourism is not as private as the situations noted above, when looking for emotional responses and reflections to certain experiences, I do not have the tools to glean this from others. What usually happens is, when we attempt to explain what motivations underlie tourist behaviour, we risk superimposing our own emotional responses onto the actions of others. It also raises ethical concerns when one is interpreting another's deep emotions.

However, this is important in this world of tourism 'experiences', so we need to understand such deep emotions. The only way that I can achieve this is to expose my own self to the eye of the reader, which is the key to autoethnography.

From Australia's Muse to a Tourism Narrative: A.B. Patterson

Towards the end of the 19th century, Australians became conscious of the bush ethos as a symbol of nationalism, as had already occurred as an unconscious process in unrecorded folklore (Ward, 1966). Of all the creative arts, late 19th century/early 20th century popular literature was highly influential as it reached the greatest number of people at the time, reflecting and influencing the tenor of the day. The popularity of the weekly magazine *The Bulletin* helped make this popular literature the most singularly influential medium of its time. The support of the writings of poets and authors such as Lawson, Paterson, Joseph Furphy and others contributed significantly to the development of an image of the Australian as 'bushman or nomadic pastoral worker, genial misogynist, laconic man of action, as Mate' (Wallace-Crabbe, 1971: xi).

Popular literature played a major role in the formation and promotion of this image, with the city-based readers eagerly supporting and reading bush literature, seeing in the bush romantic, pastoral themes removed from their place of daily routine. J.F. Archibald, co-founder and editor of *The Bulletin*, was instrumental in developing this Australian image, providing democratic and radical writers of the time with a national forum to which Henry Lawson and A.B. 'Banjo' Paterson were regular contributors (Moore, 1962).

From its early days in 1880, *The Bulletin* soon became Australia's national paper with a strong political, literary and cultural influence, expressing the exuberant nationalism of the times, particularly from the mid 1890s to early Federation. Its formation is seen as one of the most important happenings in Australian literary history, as it incorporated the democratic, brash attitudes of the intellectual mores of the day (Hadgraft, 1963). As such, any study of the development of an Australian image must be linked to *The Bulletin*.

In 1886, after submitting some verses to *The Bulletin*, Paterson met Archibald, who told him to 'have a go at the bush. Don't write anything like other people if you can help it' (Stone, 1997: 13). Paterson linked bush ballads with the popular old bush songs that were seen as the contemporary voice of the Australian people. The Australian writers who were looking for a distinctively Australian style turned from the European-style cities to the outback, founding a bush tradition that *The Bulletin* successfully turned into Legend (Crawford, 1960).

Along with portraying this recurring image of the resourceful bushman, his horse emerged as a critical element in establishing a lasting Australian image. Among the most enduring bush ballads are Paterson's 'Man from Snowy River', 'The Geebung Polo Club', 'Clancy of the Overflow' as well as Lawson's 'Reedy River' and 'Andy's Gone with Cattle', all of which incorporate 'the horse' as an integral part of the story. Many of these poems reinforced the early Australians' respect for horsemanship and daring, and there were few Australians who did not thrill at the story of the ride in 'The Man from Snowy River', which still moves me today:

> He sent the flint-stones flying, but the pony kept his feet,
> He cleared the fallen timber in his stride,
> And the man from Snowy River never shifted in his seat –
> It was grand to see that mountain horseman ride ...
>
> 'til they halted, cowed and beaten; then he turned their heads for home,
> And alone and unassisted brought them back. (A.B. Paterson, excerpt from 'The Man from Snowy River')

The 1890s publication of 'The Man from Snowy River' had the widest circulation and popular influence of all poetry and literature of the time (Palmer, 1971). Ward has also noted that the children of modern Australia's migrants seem to have adopted the 'Man from Snowy River' legend as part of their Australian identity (Ward, 1963). In fact, on a recent current affairs report by the Australian Broadcasting Commission about the wild brumbies in Australia (which comprise the central theme of the poem), the poem was shown on TV being recited around an evening campfire (Morris, 2018).

Multiple Media of *The Man from Snowy River*

As one of the three components of contents tourism noted earlier, this poem has indeed become part of contents tourism. Additional forms of content inspired by the poem, that in turn inspired tourism, include movies (*The Man from Snowy River*, 1982 and 1988), TV series (*Snowy*

River: the McGregor Saga, 1993–1996) and festivals such as The Man from Snowy River Bush Festival which has been going for over 20 years, and is still gaining in popularity (Beeton, 2015, 2016).

The movie

The emotion, story and images presented in the movie *The Man from Snowy River* and its sequel were so powerful and popular that many wished to experience this land and emotion themselves, often seeing the Victorian High Country as a 'final frontier'. Consequently, a horseback adventure tourism industry developed – prior to the movie there were two operators, while some five years after the release of the second movie, there were close to 30 (Beeton, 2001). In fact, in the early 1990s I toured with over 20 of these operators when researching the guidebook, *Beeton's Guide to Adventure Horseriding* (1994), all of whom used *the Man from Snowy River* imagery and language in their promotional material (Beeton, 2015). It was this movie and the subsequent horseback tours that got me involved in tourism as a guide on these trips, and then ultimately as a researcher and academic, in effect changing the course of my life.

When I undertook a sold-out commercial tour of the sites from the movie with the lead actor, Tom Burlinson, in 2012 (some 30 years after the movie was released, at which time many of the guests had not even been born), the guests took great delight in re-enacting iconic scenes with the star and humming the theme tune from the movie as they cantered up the mountain-side (Figure 13.1). This tour was also featured on Australian morning television news over three mornings during the six-day ride, demonstrating the ongoing power of the poem and the movie (Beeton, 2015).

Even though I had ridden that country many times before, to be there with the actor who had played such an iconic role in a movie that had changed my life was very powerful and a great thrill. To hear him recite the poem around the evening campfire, under the stars, was truly moving.

TV series

While not directly related to the poem, in the 1990s a television series was developed based 25 years on from the famous ride depicted in the poem and film, set in the fictional town of Paterson's Ridge (named for the poet, Banjo Paterson). The series, *Snowy River: The McGregor Saga* (1993–1996), was primarily made for the US market and rated well, starring many actors who have gone on to become international stars, including Guy Pearce and Hugh Jackman. A film set was built close to Melbourne at a place called Kattemingga (Figure 13.2), which stood until a few years ago when the land was sold to become a Buddhist monastery. After visiting there some years ago, at the time I believed that the owners had missed a strong tourism opportunity (Beeton, 2015).

Figure 13.1 Sharing *The Man from Snowy River* with the 'man', Tom Burlinson. Author's photo

Figure 13.2 The film set at Kattemingga. Author's photo

The festival

In northeastern Victoria, the small mountain town of Corryong claims to be the home and final resting place of Jack Riley, the 'real' man from Snowy River on whom Paterson based his 'man'. While this has not been proven, with Paterson himself claiming it was an amalgam of people,

Figure 13.3 'I was there' – tourists taking photos of themselves at the Man from Snowy River statue. Author's photo

the town thrives on the legend created by the poem (and movie), celebrating this connection with an annual festival. The Man from Snowy River Bush Festival (https://www.bushfestival.com.au) commenced in 1995 after the movie brought the story to the fore of the Australian and international imagination, attracting visitors from around the region to four days of competitions, reenactments, bush poetry and other events based around the skills celebrated in the poem and movie. While it attracts primarily domestic tourists, many of whom camp out on the recreation oval, it has reinvigorated the town, presenting another form of poetic media. The festival remains popular, featuring on ABC TV in Australia in 2017 (ABC, 2017) and replayed in 2018.

In addition to Riley's grave site, Corryong boasts a Man from Snowy River museum and a main street monument celebrating that famous (fictional) ride, which tourists use as a touchstone, taking many photos of the emotive monument (Figure 13.3).

What *the Man from Snowy River* Poem Means to Me …

I will now consider what this poem, its various media forms and concomitant tourism mean to me in a personal autoethnographic reflection; yet as I have already noted this is not an easy approach to take as such reflection can verge on the self-indulgent, and not present us with any real knowledge. However, when we consider my reflections within the contents tourism framework, I believe we can begin to understand the intimate relationships tourists have with various forms of creative content.

As I have already noted, after many years of not knowing what I wanted to do with my life, the movie and the tourism that it generated catapulted me professionally into tourism, and subsequent academic studies and research, much of which have revolved around horseback tourism and film-induced tourism. I certainly had no intention of gaining a PhD in the field, and would have laughed at the idea when I was younger, believing that I was not cut out for academic endeavour after dropping out of university at 20.

As well as setting me on such a professional trajectory, it is the emotional content of the poem and subsequent media that continues to move me. I have already mentioned how the words of the poem stir me, but after the movie was released in 1982, the music of the soundtrack continues to transport me not only to the sites and actions of the movie, but also back to the poem, and the longing and joy I felt as a child as I gazed out of the schoolroom window to the beckoning gum trees waving in the summer breeze. For some reason, my memories of the poem are always in summer.

Then, many years later, to be able to ride through that countryside on an adventure horseback tour on the horses that were used in the movie was beyond my wildest dream, as it was for many others. I lost count of the number of times I witnessed people on the horseback tours trying to emulate the wild riding epitomized in the poem and movie, even if they had never ridden a horse before (Figure 13.4).

I even participated in some bush racing based around the skills celebrated in the movie and at the festival as well as writing and presenting my own bush poetry about the horses and characters I was meeting. So, one poem culminated in my own epic bush poems.

Figure 13.4 Tourists as mountain cattlemen. Author's photo

Once I became sufficiently well known in my field of tourism research, I began to receive invitations to speak at many film and tourism related conferences, often as a keynote. In such forums, I introduced myself by showing the clip from the movie of the exhilarating ride down the mountain, accompanied by equally exhilarating music – in spite of viewing this clip hundreds of times now, I still shake with emotion at the end. Such an emotional response has encouraged others to locate the film and watch it as well as to participate in the many spin-off opportunities. Certainly, the images and music reflect the poem and also my own personal journey, as a tourist, tour guide and academic researcher.

I feel confident in claiming that the multiple media forms of the poem informed not only my personal experience, but also that of many others, creating a strong desire for them as tourists to share in such an emotional journey and re-create this for themselves, be they Australians seeking an identity or international tourists looking for that final frontier. This clearly falls into the realm of contents tourism.

Japan's Early Travel Writer and Pilgrim: Basho Matsuo

One of Japan's most revered poets, Basho Matsuo (1644–1694), was also famous during his lifetime, with many students and disciples. According to one of his followers, he lived simply, with only a rice bowl, vegetable knife and rice container (McBride, n.d.). He would take off from time to time on extended journeys across Japan which he wrote about after those journeys, leaving us with a series of travelogues, including *Oku-no-Hosomichi*, translated as *Narrow Road to the Interior* (or more commonly known in English as *Narrow Road to the Deep North*), which is the focus here.

His dominant form of poetry, haiku, is composed of 17 syllables in three sections of 5–7–5 syllables. Of course, these are in Japanese, so the translations we see do not follow this syllabic metre, but are nevertheless short jewels of observation and emotion. However, Basho Matsuo did not purely write singular haiku, but often wrote the opening verse (known as *hokku*) to a series of linked poems which were added to by other poets, as well as developing a highly emotive narrative style that combines short prose passages with haiku, known as *haibun* (Hamill, 1998). In fact, much of this prose contains subtlety of meaning and emotion along with the poems. For example, when he was lost for words by the beauty of a place, unable to write any poetry, he described it in prose when he saw the islands of Matsushima Bay (Figure 13.5):

> All covered with deep green pines shaped by salty winds, trained into sea-wind bonsai … Whose word or brush could adequately describe? (Matsuo, 1998, trans. Hamill, 1998: 16–17)

Figure 13.5 A gem of Matsushima Bay. Author's photo

Basho's *haibun* was originally presented in a written format in scrolls written and published after his journeys, while his haiku was written and at times presented to his hosts while on the road. Since that time, his work has been published in translated books, each presenting us with quite different experiences, which can often be the case when reading translations. In fact, even translating his Japanese into modern Japanese is fraught, let alone to other languages such as English. In effect, they are all multiple forms of media representations of his work.

Basho's walks were more than simple peregrinations, often following the paths of those he revered as well as seeking out sites of powerful historical and spiritual meaning. His own trips were very personal pilgrimages, reflecting on an ancient, sacred past, which in turn created an itinerary that others still follow today.

There are many who have written about their own pilgrimage, where they walked in Basho's footsteps, such as Lesley Downer, who presented this very much as her own adventure at a time when few Westerners were taking these trails (Downer, 1989). While I have read her account and can sense her own passion, I was not entirely engaged with this work, possibly as I read it prior to my own experience. Furthermore, she found a Japan that was barely recognizable, yet I found Basho's Japan behind the

Figure 13.6 Basho Monument on the trail near Hiraizumi. Author's photo

landscape, within his writings and in no small part owing to the curation of the tour that I participated in. After my trip I was given access to the work of scholar John McBride who researched (and walked) Basho's journey for over 28 years before handing over to the company I did my trip with, Walk Japan (McBride, n.d.). Along with Hamill's (1998) translation, I found this to be work that I connected with and that tied me in to Japan and Basho, alongside the tour. Other media representations of *Narrow Road to the Interior* that I experienced while travelling included a short play along with numerous statues, monuments and museums dedicated to Basho (Figure 13.6a and 13.6b).

Finally, there is the actual specialized tour in which I participated, which in effect is another form of 'media' as well as being based on the media of Basho and those who re-presented his work.

My Basho Autoethnographic Experience: The Tour

As noted earlier, Basho's own journeys were highly personal pilgrimages, travelling in the footsteps of others he admired, so at times we were thinking about and connecting with people and incidents over a thousand years old, not only his own journey. Consequently, our experience felt at times like time-travelling, moving in and out of these hundreds of years. The emotive nature of poetry and Basho's poetic prose brought us not only to Basho's time but also to those places and activities whose past he cherished so.

The beginning sentences and haiku of *Narrow Road to the Interior* set us on our way, and resonated with me with its references to horses and the journey of life:

> The moon and sun are eternal travelers. Even the years wander on. A lifetime adrift in a boat, or in old age leading a tired horse into the years, every day is a journey …
>
> Spring passes
> And the birds cry out – tears
> In the eyes of fishes. (Matsuo, 1998, trans. Hamill, 1998: 3–4)

The contents of the tour provided me with the opportunity to connect in a deep way with many of the places I visited, resulting in a highly emotional and moving, spiritual experience. Along the way we not only read Basho's work, but visited monuments, viewed re-enactments and shared our experience on social media, at times re-creating famous scenes. Some even wrote their own haiku, but I did not feel bold enough to do that. In fact, I was overwhelmed by the depth of meaning that Basho was able to put into a few syllables, although, now that I am back home in Australia, I am trying to express the beauty of my seaside home via the delicate touch of haiku, with one attempt below:

> Bird wingtips flashing
> Diamond cut water ripples
> The turning begins

I do not intend to relate this tour via the geographic itinerary we followed, which is outlined in Figure 13.7, but more through my personal emotions, thoughts and themes. However, my first Basho site, Urami Falls, hit me hard – it was very exciting, and I felt an immediate connection with the past, being instantly transported to a time over 200 years ago. I also felt a sense of wonder and joy at being there and being able to feel Basho's joy through his own words. In Basho's time, there was a path that went behind the falls where there was a small cave. Today the path has collapsed, but it is still easy to picture the scene. He sat there and wrote:

> Stopped awhile
> Inside a waterfall –
> Summer retreat begins. (Matsuo, 1998, trans. Hamill, 1998: 6)

Just as this marked the beginning of our trip, so it was for Basho ('summer retreat begins'), and I shared my video of the grasses swaying in the air movement that was created by the waterfall on social media. It was absolutely delicate yet powerful in an intimate way (Figures 13.8a and 13.8b).

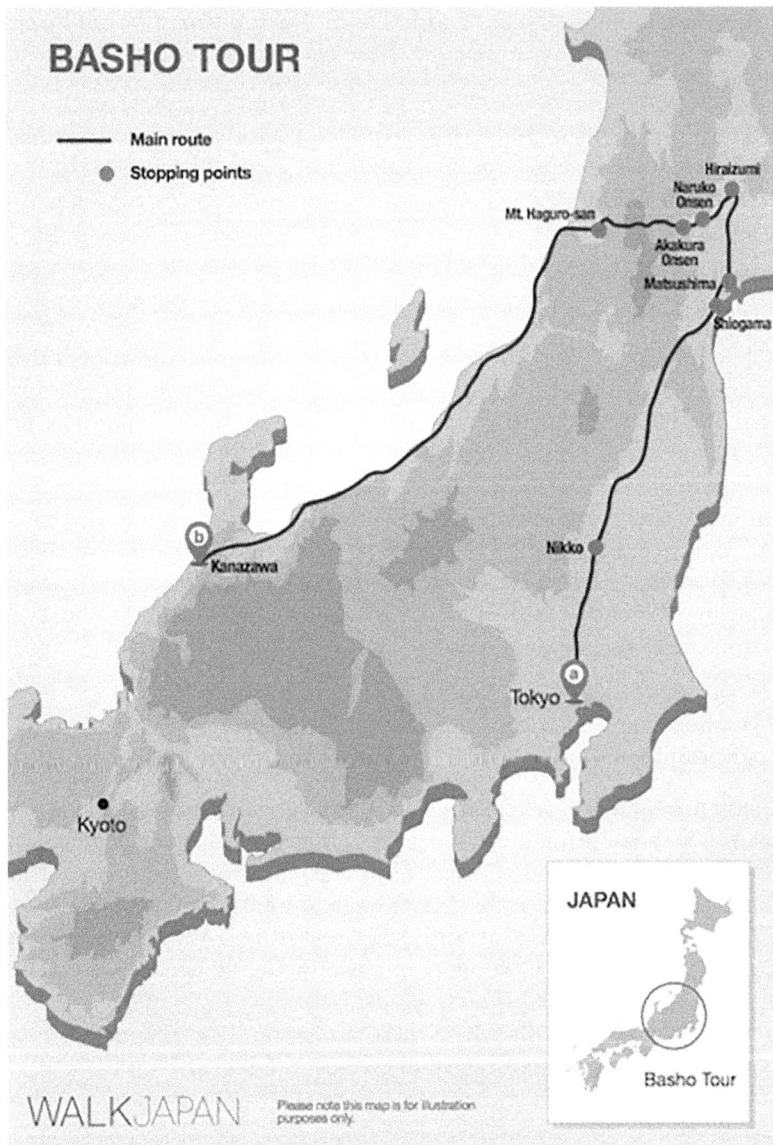

Figure 13.7 Basho Tour itinerary (source: Walk Japan Tour Notes)

Moving from the simple yet profound beauty of nature, UNESCO World Heritage listed Hiraizumi is the site of a great battle of the Warring States era from the end of the 15th century to the end of the 16th century, some time even before Basho. All that was left even by the time he visited was a field and ruins. Today, it remains a lonely empty field with only some foundations of the pillars visible.

Figure 13.8 Urami Falls that Basho sat behind. Author's photo

> Summer grasses:
> all that remains of great soldiers'
> imperial dreams. (Matsuo, 1998, trans. Hamill, 1998: 19)

This poem and the empty field left me feeling pensive and sad, and Basho's words struck me as to the futility of such battles, along with our own hubris – we all believe we (and our causes) are so important, yet in the end all that is left is nature. Reading the haiku now brings it all back to me, yet it is not a bad or sad feeling, but very deep and contemplative.

Basho was lost for words at the sacred mountain of Haguro-san, where we climbed 2466 steps flanked by giant 600-year-old cedars and small shrines up to the main Shinto Shrine. We stayed in the Shrine complex with pilgrims (as Basho did) and participated silently in morning prayers, as did Basho:

> To say more is sacrilege. Forbidden to speak, put down the brush, respect Shinto rites. (Matsuo, 1998, trans. Hamill, 1998: 25)

The next morning before prayers I walked around the complex, taking in all of the individual shrines and many Jizo – small monk-like statues that protect children (including the unborn who died), women and travellers, which I love as, for me this reflects the way we should travel – with child-like curiosity.

Even here Basho is revered and immortalized in a monument set in a prominent position near the shrine (Figure 13.9).

Figure 13.9 Basho monument at Haguro-san, where he was too moved to write. Author's photo

One of many highlights of our walk in Basho's footsteps was the Border Guardhouse at Sakaida, which is the only existing building Basho visited, and one where he spent a night. What was initially a rather odd haiku made sense once we visited and sat, possibly on the same wooden platform that Basho did. The guardhouse was also where horses were bred and housed, along with the family and travellers. Basho clearly did not have a great time there as his haiku tells us in no uncertain terms:

Eaten alive by
lice and fleas – now the horse
beside my pillow pees. (Matsuo, 1998, trans. Hamill, 1998: 20)

Once I realized that the horses were in the same dwelling, and that the usual translation of 'the horse' as a singular creature peeing was not necessarily correct, owing to there being no plural in the language of the time, it became clear to me why he was so distressed. As a person who has spent some time with horses (and slept with them), I am aware that when one urinates, most of the others follow suit. Consequently, this may have been quite a loud and smelly event! I felt rather privileged to have this personal insight into one of Basho's most famous haiku.

At the onsen (hot spring) we stayed in near Sakaida, we were treated to a re-enactment of his stay at the Guardhouse (without the smell of horses!) and ate a typical breakfast that Basho would have consumed,

Figure 13.10 Local residents re-creating Basho penning a haiku after his night at the Guardhouse. Author's photo

known as *Nara Chazen*, which consists of four red bowls on a black lacquer tray, comprising sweet black beans, pickles, rice with chestnuts and mountain vegetables (McBride, n.d.). The re-enactment was charming and low-keyed, being presented while our small group of 12 had dinner, and was presented entirely in Japanese. However, we knew the story (Figure 13.10).

After my experience at the guardhouse and onsen, I began to feel connected to this famous poet and the people who revere him so.

What Basho's Haiku Mean to me …

I have presented some autoethnographic reflections while describing aspects of my tour in Basho's footsteps, so I will not repeat myself here, but wish to explain how I now feel regarding Japan in light of my experience, following on from the previous sentence.

First, it is important that we look at the various translations of Basho's poetry into English (as well as modern Japanese) as separate media forms owing to the variations and emotions they present. While many English translations refer to this work as *Narrow Road to the Deep North*, I chose to rely primarily on the translation by Sam Hamill, not in the least for his explanation of titling his translation as *Narrow Road to the Interior*:

> His journey is a pilgrimage; it is a journey into the interior of the self as much as a travelogue, a vision quest that concludes in insight ... The journey itself is home. (Hamill, 1998: xx)

It is not easy to get under the skin of a culture that hides behind politeness and rules, as in Japan, yet this trip gave me just that. Not only was I moved while on this journey, but I found that my knowledge of one of Japan's great poets, and my experience of walking in his footsteps, gave me an entree into Japan that I had not received in spite of many visits over the previous years. I now feel strong emotional connection to the country, its nature, culture, religions and people, and a certain pride that I know a little about one of their most revered poets in a land that truly treasures such art.

Conclusion: The Power of Poetry in Tourism

When we think of 'popular culture' our thoughts tend towards current media trends, such as film, TV, YouTube formats, music and social media, yet poetry remains one of the most powerful emotional connectors for many of us.

As I noted in the early sections of this chapter, I have examined my own deep connections with the poetry of my homeland and how it affects my domestic travel experience alongside the poetry of another place and culture and how it has deepened my international travel experience. I believe that the two quite distinct poetic forms, in structure, time and distance, and the way that they have informed my tourist experiences are similar in their highly emotive content and that they have enabled me to connect deeply with elements of my own and others' cultures.

To be honest, I did not expect that my experience in Japan with Basho would prove to be as powerful, demonstrating the importance of self-reflection in tourism and our research.

I trust that I have succeeded here in uncovering the power of poetry as contents tourism, by exploring the poetry of my homeland and its effect on my domestic travel experience alongside the poetry of another place and culture on my international experience.

References

ABC (2017) Back Roads: November 27. See https://tvtonight.com.au/2017/11/back-roads-nov-27.html (accessed November 2018).

Beeton, S. (1994) *Beeton's Guide to Adventure Horse Riding*. Balwyn: On Track Tourism Consultants.

Beeton, S. (2001) Horseback tourism in Victoria: Proactive crisis management. *Current Issues in Tourism* 4 (5), 403–421.

Beeton, S. (2008) From the screen to the field: The influence of film on tourism and recreation. *Tourism Recreation Research* 33 (1), 39–47.

Beeton, S. (2015) *Travel, Tourism and the Moving Image*. Bristol: Channel View Publications.

Beeton, S. (2016) *Film-Induced Tourism* (2nd edn). Bristol: Channel View Publications.

Beeton, S., Yamamura, T. and Seaton, P. (2013) The mediatisation of culture: Japanese contents tourism and pop culture. In J.-A. Lester and C. Scarles (eds) *Mediating the Tourist Experience: From Brochures to Virtual Encounters* (pp. 139–154). Farnham: Ashgate.

Crawford, R.M. (1960) The birthplace of a culture. In C. Wallace-Crabbe (ed.) *The Australian Nationalists: Modern Critical Essays* (pp. 220–224). Melbourne: Oxford University Press.

Denzin, N.K. (1997) *Interpretive Ethnography: Ethnographic Practices for the 21st Century*. Thousand Oaks, CA: Sage.

Downer (1989) *On the Narrow Road to the Deep North: Journey into a Lost Japan*. London: Jonathan Cape.

Hadgraft, C. (1963) Literature. In A.L. McLeod (ed.) *The Pattern of Australian Culture* (pp. 42–101). Ithaca, NY: Cornell University Press.

Hamill, S. (1998) Translator's introduction. In B. Matsuo *Narrow Road to the Interior, and Other Writings* (trans. S. Hamill) (pp. ix–xxxi). Boulder, CO: Shambala.

Matsuo, B. (1998) *Narrow Road to the Interior, and Other Writings* (trans. S. Hamill). Boulder, CO: Shambala. (Oringinally published 1688.)

McBride, J. (n.d.) The narrow road of Oku: Matsuo Basho's Oku no Hosomichi (unpublished manuscript).

Moore, T.I. (1962) The red page rhadamanthus: A.G. Stephens. In J. Jones (ed.) *Image of Australia* (*The Texas Quarterly*, summer edn) (pp. 96–103). Austin TX: University of Texas.

Morris, N. (2018) Record attempt to train 70 brumbies in seven days transforms horses and humans. ABC 7.30 Report, 19 February. http://www.abc.net.au/news/2018-01-30/training-70-horses-in-7-days-tranforms-horses-and-humans/9359244 (accessed September, 2017).

Palmer, V. (1971) The legend. In C. Wallace-Crabbe (ed.) *The Australian Nationalists – Modern Critical Essays* (pp. 1–21). Melbourne: Oxford University Press.

Reijnders, S. (2011) *Places of the Imagination: Media, Tourism, Culture*. Farnham: Ashgate.

Stone, W. (1977) *The Best of Banjo Paterson*. Sydney: Lansdowne Press.

Wallace-Crabbe, C. (1971) Introduction. In C. Wallace-Crabbe (ed.) *The Australian Nationalists – Modern Critical Essays* (pp. ix–xiii). Melbourne: Oxford University Press.

Ward, R. (1963) The social fabric. In A.L. McLeod (ed.) *The Pattern of Australian Culture* (pp. 12–41). Ithaca, NY: Cornell University Press.

Ward, R. (1966) *The Australian Legend* (2nd edn). Melbourne: Oxford University Press.

Conclusions: Sustainable Contents Tourism in the 21st Century

Philip Seaton

In the 13 chapters of this book, the authors have presented examples of contents tourism with an international dimension. By expanding the application of contents tourism theory and method to international tourism, our main aim has been to break out of the associations between contents tourism and Japan (although Japan has maintained a strong presence in this book), and to demonstrate the concept's worldwide usefulness and applicability. This was the task we left ourselves as a research team at the end of *Contents Tourism in Japan* (Seaton *et al.*, 2017: 267), which had situated contents tourism firmly within the Japanese context. In this concluding chapter, I would also like to extend the remarks beyond the scope of simply this book and consider the main conclusions of our project (funded by a Japan Society for the Promotion of Science Grant, 2014–2019), which has involved dozens of researchers from East and Southeast Asia, North America, Europe and Oceania.

As discussed in Takayoshi Yamamura's introduction, contents tourism discourse emerged in Japan in the 2000s and entered the language of official government policy in 2005. After the key early works in Japanese, particularly by Masubuchi (2010), Yamamura (2011), Okamoto (2013) and others, the term contents tourism entered the English-language literature in 2013 (Beeton *et al.*, 2013). It has been developed since then through a number of key publications: a special edition of the journal *Japan Forum* (2015, Volume 27.1), discussion within the second edition of Sue Beeton's (2016) seminal *Film-Induced Tourism* (2nd edition), the monograph *Contents Tourism in Japan* (Seaton *et al.*, 2017), a special edition of the *Journal of War & Culture Studies* (Volume 12.1, 2019) and the present edited volume. There have been many other articles and presentations by project members in English, which are listed in full on the project's grant page within the Japan Society for the Promotion of Science website (JSPS, 2014). Of course, research also continues in Japan in Japanese, although the Academy of Contents Tourism, Japanese Association for Contents

History Studies and other groups and individuals have yet to make a concerted effort to publish their research findings for an international audience. At the government level, efforts to attract in-bound tourists using manga, anime and other aspects of Japanese pop culture continue apace alongside the broader Cool Japan strategy. Local authorities remain active in trying to convert local contents into tourism resources as they have been encouraged to do by central government since 2005.

Over the course of the project, our definition of contents tourism has evolved. We started with a relatively vague definition that situated contents tourism within cultural tourism as a close relative of film-induced tourism, literary tourism and other forms of media-induced tourism (Beeton *et al.*, 2013; Seaton & Yamamura, 2015). We shifted to a one-sentence definition in *Contents Tourism in Japan* (Seaton *et al.*, 2017: 3) – 'travel behavior motivated fully or partially by narratives, characters, locations, and other creative elements of popular culture forms including film, television dramas, manga, anime, novels, and computer games' – and have modified the definition further in this book to place the process of 'contentsization' at its core. To repeat the full definition from Takayoshi Yamamura's introduction:

> Contents tourism is a dynamic series of tourism practices/experiences motivated by contents (defined [...] as 'information – such as narratives, characters, locations, and other creative elements – that has been produced and edited in popular culture forms and that brings enjoyment when it is consumed'). Contents tourists access and embody 'narrative worlds' that are evolving through 'contentsization', namely the continual process of the development and expansion of the 'narrative world' through both mediatized adaptation and tourism practice.

The result is a concept fit for the digital and internet ages, in which fans are not only passive consumers of works of popular culture, but also active producers of derivative works, online discussion and fan communities. Likewise, when they travel to pursue an interest in their favourite contents, fans are not simply passive tourists consuming a pre-packaged experience. Their online and offline voices are crucial in shaping the events and experiences offered at tourist sites. On occasions, fans may even be the driving forces behind the creation of tourist sites and tourism experiences, and these experiences then contribute to the evolution of the narrative world. The concept of contentsization, therefore, focuses attention on how a narrative world created by works of mediatized popular culture is developed, transformed and expanded not simply by the original creators of the works, but also by fan communities and the communities who host related tourist sites/sights visited by those fans. These interactions between the three main players – contents businesses, local authorities and fans – are at the heart of all analysis of contents tourism.

There remain occasions, of course, when a contents tourism approach might be less appropriate. Contents tourism prioritizes the social and cultural dynamics of tourism induced by mediatized culture. As such, it is unlikely to appeal to those wanting short-term quantitative answers, such as the economic impact of a particular work (which we think is 'unknowable' in most situations, particularly over the long run). Contents tourism research focuses on longer-term issues such as the evolution of a narrative world, the ongoing relationships created between fans and communities and the ways in which communities welcome, or not, sets of contents as part of their local heritage. Furthermore, tourism phenomena associated with one-off films or visitation at an author house museum out of enjoyment simply of the author's novels remain archetypal film tourism or literary tourism experiences, respectively. When both the copyrights surrounding the contents and the tourism experience are very tightly controlled (such as at Disneyland), the opportunities for fans to engage in contentsization either in mediatized fan adaptations or in tourism practices are also restricted.

In addition to all of these limitations, it should also be noted that the various research groups using the term 'contents tourism' do not agree on either the definition of contents tourism or the parameters of contents tourism research. Our research project has excluded some topics that other groups have addressed, in particular tourism PR mascots/characters, games such as Pokémon Go that have no associated story linked to the place where the game is played, and any advertising borrowing from popular culture styles (for example, PR songs sung by local idol groups). As such, many cases fall into the 'debateable' category because they exhibit clear characteristics of contents tourism in some ways but not others. Overall, the state of contents tourism research, like the narrative worlds it analyses, is in a continual process of development, transformation and expansion.

Sustainable Contents Tourism and Policy Implications

It would be unusual for a research project of this scale not to make any comments on policy. In the English-language publications related to the project thus far, we have avoided policy implications, primarily because until we reached a critical mass of case studies from both outside and inside Japan it seemed premature. However, the time now feels right to offer conclusions for tourism practitioners based on our research results.

The contents production phase

We see contents tourism as an *unplanned* effect of works of mediatized popular culture. Works produced primarily to attract tourists are advertising or PR and are beyond the scope of our research. However, the distinction between works which do and do not have an explicit intention to induce tourism has become increasingly blurred as the potential economic benefits of contents tourism (particularly film-induced tourism) have become evident. Many local authorities and national governments now offer incentives to contents producers to make films, dramas and other works in their regions. Film commissions, arts councils, public funding bodies and other such institutions support productions, usually with at least one eye on potential tourism effects.

The economic benefits of such support exist in two phases, *during production* and *post-release*. Revenues from the production process are primarily revenues from location-hunting tours (see Yamamura, Chapter 4), hotel occupancy by actors and production staff during filming, and related temporary contracts given to local businesses such as catering and construction. Tourism by film/drama actors and production staff on their days off, or by fans keen to witness production, or people somehow connected to the production (such as the spouses of extras) might also be a factor. However, in most cases significant tourism effects only come after the release of a work. This is particularly the case for works such as manga, novels and games, where there is little of the publicity or entourage surrounding a major cinematic/dramatic production. Given the unpredictability of what contents will trigger a tourism boom (see 'What Contents?' below), a prudent policy at the production stage is to assume that support for contents production (particularly film, anime and drama production) is a self-contained project worthwhile for its immediate economic benefits and intangible benefits for the community, for example, local residents saying, 'It was an unforgettable experience being an extra in a film made in our town'.

Preparing for the post-release phase

Local authorities and businesses should ensure they are ready to accept an influx of tourists post release, but at the same time have realistic expectations regarding how many people will actually travel as contents tourists. Treating production phase support as a loss-leading endeavour in preparation for an anticipated post-release tourism boom carries significant financial risks. The sets of contents that trigger massive tourism phenomena are the exception rather than the rule. While examples such as *Lord of the Rings* tourism in New Zealand might raise expectations, in reality the vast majority of contents tourism booms are modest, temporary windfalls, and not life-changing events for local communities, especially if contents tourism is supplementing a pre-existing tourism industry. Many local authorities and tourism operators have incurred significant losses betting on contents tourism booms that never materialized (Seaton et al., 2017: 157–159). Consequently, if the local authority or tourism businesses are determined to invest in cultivating contents tourism, keeping investment levels within the scope of 'acceptable losses' and negotiating licensing agreements with copyright holders during production are essential (Seaton, 2019a, 2019b). Contents tourism booms typically have a short lifespan (the public's attention span is short) and starting negotiations with copyright holders after the boom has started can be closing the barn door after the horse has bolted. The other approach is to view investing in contents tourism as the long-term cultivation of cultural heritage, or as Christopher Hood discusses in Chapter 11, the cultivation of corporate image and the corporation as fan.

One caveat here is that there is a particular type of fan who does not want touristification. Instead, such fans want to blaze an unbeaten trail in which subsequent waves of fans will follow (Okamoto, 2015). These fans are not interested in arriving at a location only to find tourist maps for the masses, pre-packaged experiences and endless licensed souvenirs. Stefanie Benjamin's autoethnography (Chapter 12) has spoken directly of this fan desire not to be made to feel like a potential consumer ripe for commercial picking. On this point, the Hobbiton attraction in New Zealand is a revealing case study (Seaton, 2019b). There were two main phases in the development of Hobbiton as a film tourism attraction. The first was after the release of the *Lord of the Rings* trilogy (2001–2003), when visiting the remains of the dismantled set proved to be a popular film location tourism experience for fans. The second phase was after the release of *The Hobbit* trilogy (2012–2014), when an agreement concluded before filming between the landowners and production company meant that the set was preserved for use as a film tourist attraction (Figure 14.1). Today, Hobbiton is a

Figure 14.1 Picturesque Hobbiton. Author's photo

beautifully kept attraction in a very picturesque location, but when a site of film or contents tourism evolves into a packaged tourism experience, it can leave fans feeling alienated. Contrasting her 2014 visit with earlier ones in 2003 and 2006, Sue Beeton (2016: 128) comments, 'For me there was no longer any "movie magic" there, simply a theme park'.

Nevertheless, the incredible commercial success of Hobbiton (around 3000 visitors a day, adult entry NZ$84.00 in early 2019) does offer a useful blueprint for commercially successful touristification. If the first wave of fans visiting related sites offers evidence of a potential boom, then local authorities and businesses should have ready-made plans (particularly regarding licensing rights) to increase quickly the provision of services and merchandise, first for pioneer fans, then second-wave fans who base their travel on the word-of-mouth emanating from pioneer fan communities, and finally the mass tourism triggered by broader attention in mainstream media. However, local authorities and businesses should always be aware of the trade-off between the commercial gains of appealing to a mass audience and the risk of over-commercialization alienating their dedicated fan base. If fan satisfaction is a primary consideration, then modest levels of tourism and modest financial returns for local authorities and tourism businesses might be the levels of economic activity at which the quality of the tourism experience is optimized for fans.

What contents?

What works trigger tourism in the post-release phase is the source of much speculation and uncertainty. While there are success stories from most genres, from crime thrillers to space fantasies, equally there are works in all genres that do not trigger tourism. There is no magic formula for a work that triggers contents tourism. The nearest our research project has discovered to a 'sure bet' that consistently triggers tourism is the annual Taiga Drama (Seaton, 2015) and Morning Drama (Scherer & Thelen, 2017) on Japanese television. The keys here are that: (a) they are an annual television institution with a half-century of history (these drama series began in the 1960s); (b) the drama runs for six months (morning drama) or 12 months (Taiga Drama), giving ample time for people to engage; (c) there is massive associated media attention, including in the pre-broadcast phase, such as guest appearances on other programmes by the leading actors; and (d) the narratives are rooted in specific locations, so the tourism destination is clearly identified from the start. Even so, there is no sure formula for predicting which Taiga Dramas and Morning Dramas will gain high viewing figures, or to what extent viewing figures will convert into increased tourism. It is not necessarily the case that the most popular dramas precipitate the largest levels of tourism or the greatest financial effects. In addition to the quality and resonance of the contents, tourism levels depend on a complex range of factors, such as destination accessibility, the type of tourism that may be engaged in (walking tours at locations generate far less revenue than boat tours, for example), the capacity of the destination to accommodate visitors (particularly overnight hotel stays), the broader appeal of the tourism destination beyond contents tourism and external factors such as macroeconomic conditions, natural disasters and even the weather.

We have identified three particular types of contents that over the course of many works and across various media platforms are good at building up a fandom of the 'narrative world' that induces tourism. The first type is actual history and traditional stories that trigger tourism to related heritage sites, a phenomenon we have termed 'heritage and/or contents tourism' (Seaton *et al.*, 2017: 32–33). The original 'authorship' of such tales lies with the historical figures – whether real, mythical or somewhere in between – whose deeds have inspired others to (re)tell their stories down the ages. In such cases, tourists tend to visit heritage sites (Frost, 2006: 253) and there can be considerable synergies between heritage, screen and literary tourism as noted by Agarwal and Shaw (2018). The nearest example in this volume is the yōkai tourism described by Shinobu Myoki in Chapter 6.[1] The second type is based on 'canonized contents', namely fictional stories (particularly those first disseminated via novels and the written word) that generate many adaptations over time. Typical examples are the narrative worlds around classical literature, such as those featured in part one of this volume: Jane Austen (Seaton, Chapter 1), the Brontës (Thyne and Larsen, Chapter 2) and *Heidi* (Yamamura, Chapter 4). Sue Beeton (Chapter 13) also demonstrated the potential of poetry. The third type is the contemporary pop culture franchise, for which the copyright holder determines the terms of contents dissemination (Jaworowicz-Zimny, Chapter 3; Jang, Chapter 7; Sugawa-Shimada, Chapter 8). The main lesson for tourism practitioners is that a reasonable proxy for tourism potential is the ability of the narrative world of the original work to develop in multiple subsequent works across various media formats. We have positioned tourist sites as 'media', so the emergence of tourist sites falls within the scope of the media mix and contentsization of transmedia franchises. In this context, popular historical figures (historical contents) or series of novels that spawn multiple successful screen and other adaptations typically indicate the presence of a resonant 'narrative world' with strong contents tourism potential. Furthermore, the more that works of 'pop culture' seem to be transforming into 'literature', 'classics' or 'heritage', the more likely it is that any related tourism phenomenon is going to share the longevity of the contents.

The booms associated with one-off works or 'flash in the pan pop culture', in contrast, tend to be small and short-lived. Examples of one-off films that generate sustained tourism are very rare. Perhaps the best example here is *The Sound of Music* and associated tourism in Salzburg (Im &

Chon, 2008). However, even this example might be better considered as contents tourism rather than film-induced tourism because much of the film's enduring popularity lies in the widespread performance of its songs. This indicates the potential of the contents tourism approach for studying Bollywood tourism, where the 'filmy songs' often become stand-alone hits in their own right (Nanjangud, 2019). Many other one-off films find their tourism potential achieved in theme parks, such as Disneyland and Universal Studios. However, the rides and attractions at theme parks, along with their cinematic or other themes, must be renewed regularly in order to attract repeat visitors. Furthermore, the clear evidence from Universal Studios Japan (Beeton & Seaton, 2018) is that innovative management of the theme park to maximize its appeal across various demographics and to encourage repeat visitation (for example at seasonal events), and not simply interest in the original contents, is vital for sustaining tourism levels.

What destination?

Our final major conclusion is that it may not even be the contents that are decisive in triggering the boom. It may be the nature of the destination and its intrinsic tourism potential unrelated to contents. As indicated in the surveys of Jane Austen fans in Chapter 1, while contents tourism might be a primary motivation for American fans visiting the UK, it is rarely the only one. A destination's broader touristic appeal can assume even greater significance than the appeal of the contents when people are planning trips. New Zealand's stunning scenery was attracting many tourists before it became 'Middle Earth', and as Catherine Butler described in Chapter 5, the Cotswolds have a massive appeal to Japanese tourists beyond providing the locations for a number of manga and anime. Indeed, such contents may have had far less commercial (let alone tourism) potential if they were set in little-known, uninteresting or unattractive locations. Even if the contents tourism experience is the primary motivation for visiting a destination, time not engaged directly in contents tourism still has to be filled with something that makes the travel worthwhile overall. It is the particularly dedicated fan who selects a tourism experience where the entire experience is about engagement with (Benjamin, Chapter 12) or performing (Rastati, Chapter 9) the contents. For most tourism experiences, contents tourism is just one of many components of an attractive touristic experience, although a set of contents can be the main reason for, or even the core theme running through, an entire trip to a foreign country (Kim, Chapter 10; Beeton, Chapter 13).

From the perspective of the contents tourism researcher, the nature of the destination also affects the viability of the research project. Trying to find the James Bond fans hunting for locations in London in amongst the millions of other tourists is daunting, unless they drop by a dedicated attraction such as the London Film Museum. In any large metropolitan centre with a large tourism industry, aggregate tourism data is virtually useless for identifying contents tourism trends. Typically, it is the destinations away from urban centres, where the number of plausible explanations for spikes or dips in visitor numbers is limited, where the clearest evidence of contents tourism may be found. For tourism practitioners in urban areas, therefore, getting clear evidence of contents tourism is usually only possible via interviews and surveys. In contrast, at individual sites with an entrance fee and in isolated areas with little prior tourism, the changes in tourism statistics around the release of specific works of popular culture can offer clear evidence of contents tourism activity.

Summary

In short, there is no simple prescription for success in contents tourism, and there is no reliable crystal ball for predicting which works or franchises will trigger contents tourism. Consequently, there will always be gaps between expectations and reality in the levels of contents tourism, resulting in both unexpected success stories and widely publicized failures. Every case is distinctive and only reveals the keys to its success or failure over time. Rather than a definitive list of what to do to make contents tourism occur, therefore, it is more useful to identify a set of basic principles or best practices for practitioners to follow based on the lessons from past successful examples of contents tourism. These principles also tie in with the broader economic and environmental challenges facing the tourism sector in the 21st century.

Contents Tourism and 'The Doughnut'

Contents tourism is an old phenomenon. In this book we have mentioned literary tourism related to Jane Austen and the Brontës in the 19th century, and in *Contents Tourism in Japan* we traced contents tourism and its antecedents back centuries further still to the early stages of the Tokugawa Period (1603–1867) and before. However, in its modern guise, contents tourism is very much a 21st-century phenomenon whose characteristics owe much to the internet revolution, social media and online communities. As such, a 21st-century vision for contents tourism policy and practice is required.

Today, the greatest challenge facing not only the tourism industry, but also the planet as a whole, is the spectre of environmental breakdown caused by global heating, pollution and excessive resource consumption. The Holocene has made way for the Anthropocene, and our species now lives on a planetary eco-system heading into uncharted territory. Just before this book went to press, the UN Intergovernmental Panel on Climate Change warned that we had only 12 years left to act before the worst effects of climate change become irreversible (Watts, 2018). The primary historical responsibility for this situation rests with the advanced industrial nations, who account for the vast majority of greenhouse gas emissions and resource overuse. However, the travel industry, particularly in Asia, will be a key player in the equation during the struggle to limit environmental damage. As millions of people in Asia have been lifted into the middle classes by economic development in countries such as China, India and Indonesia and therefore become able to afford travel for leisure, there has been an explosion in levels of international tourism. Visitors to Japan, for example, increased from 5.2 million in 2003 to 31.2 million in 2018, with over three-quarters of visitors from Asian countries (JNTO, 2019). The environmental pressures caused by tourism look set to continue on a steep upward trend. Tourism's contribution to climate breakdown has become an important area of research and a new buzzword, 'overtourism', has emerged. The keyword for successful contents tourism, therefore, must be *sustainability*, and a way of envisioning how this may be achieved in tourism can be seen by using one of the most important emerging concepts in the field of economics: 'the doughnut'.

Doughnut economics is a bold new vision for economic activity in the 21st century. In *Doughnut Economics*, Kate Raworth roots her arguments in disillusionment with traditional economic theory and the obsession with markets, shareholder returns and GDP growth in neoliberal economics that have taken our planet to the brink of environmental breakdown. She urges us to stop thinking of the economy in terms of a machine which automatically self-corrects via the mechanisms of the market, and to start thinking in terms of a complex system sustained by and inextricably linked to the planetary system. She envisions the economy as a doughnut with a hole in the middle (Figure 14.2). The 'essence of the doughnut' is that there is 'a social foundation of well-being that no one should fall below, and an ecological ceiling of planetary pressure that we should not go beyond. Between the two lies a safe and just space for all' (Raworth, 2017: 11). The aim of humanity should be to 'get into the doughnut', which also acts as our 'twenty-first-century compass' (Raworth, 2017: 10, 43).

The doughnut concept may be adapted for contents tourism. In our model of contents tourism, the interactions between the three actors are depicted as circles and the relationships between them are indicated using arrows (Figure 14.3).

However, we can modify this version of the model to depict sustainable contents tourism. The circles are changed to doughnuts, which indicate the 'safe and just' zones in which each of the three actors operate. The doughnuts become interlocking to indicate where the respective 'safe and just spaces' or 'ideal spaces of activity' of the three actors meet. If each of the actors operates within their own doughnut, while simultaneously cooperating with other actors to ensure that they may operate within their doughnut, it should be possible for everyone to benefit while at the same

234 Contents Tourism and Pop Culture Fandom

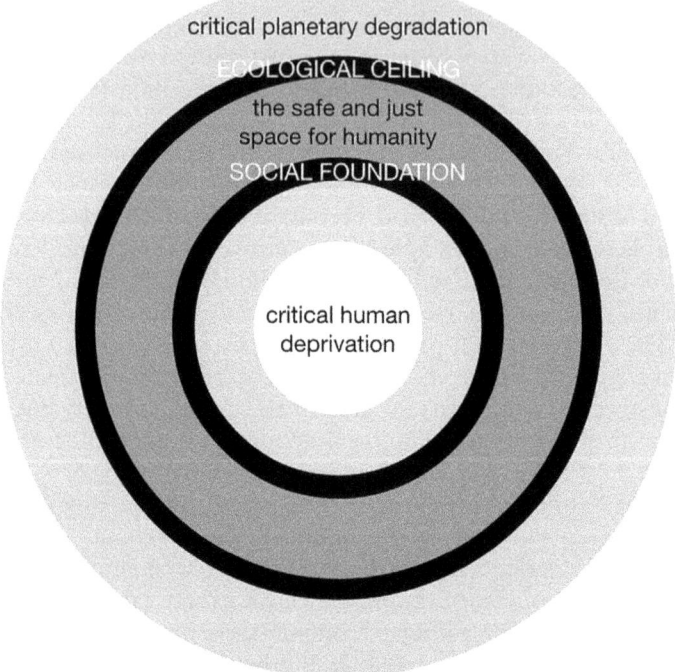

Figure 14.2 The doughnut of *Doughnut Economics*. Source: Raworth (2017: 11)

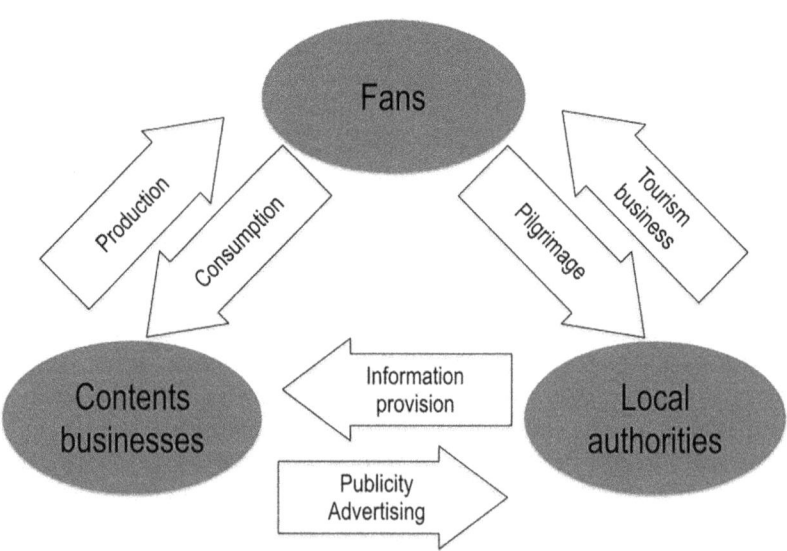

Figure 14.3 The actors of contents tourism. Source: Seaton et al. (2017: 39)

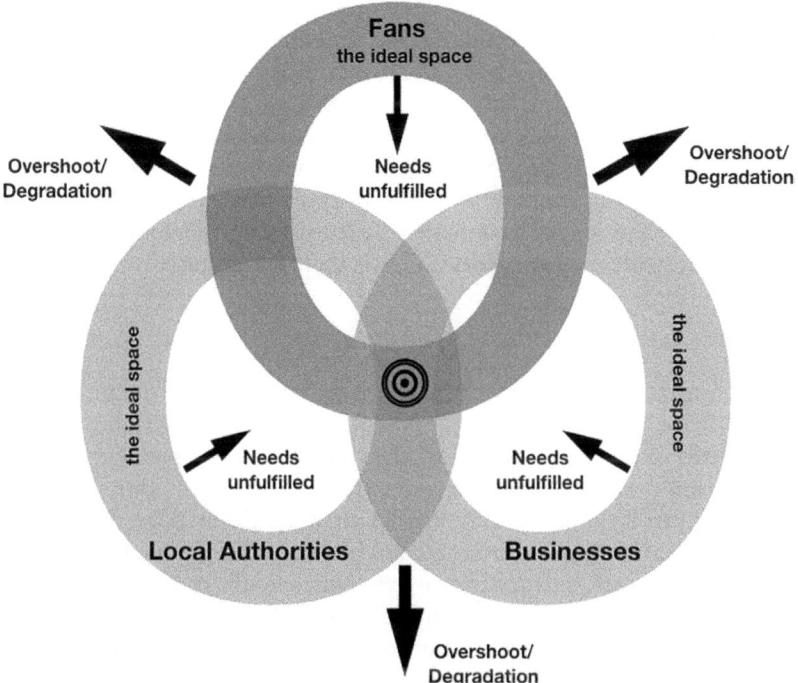

Figure 14.4 The 'safe spaces' for contents tourism. Modified by the author based on Raworth's (2017: 11, 44) concept of the doughnut

time achieving a balanced state that minimizes degradation to society and the planet. This is the target in the middle of Figure 14.4: the ideal space for contents tourism.

However, what are the foundational needs and ceilings for each of the three actors?

Local authorities

For local authorities, the keywords of their doughnut are *sustainability* and *identity*. Sustainability refers to both *economic* and *environmental* impacts of contents tourism. The community wants to attract tourists who will spend money. However, if there is overtourism then problems arise, such as disruption to community life (noise pollution, traffic congestion) or excessive pressure on the local ecosystem (littering, over-extraction of water resources). However, if tourism fails to bring sufficient economic benefits, remaining a tourist site is only possible if the community underwrites any financial losses. The issue of identity, meanwhile, refers to whether the community is happy to identify with a particular set of contents. If the community draws pride from its associations with contents, there will be support for maintaining the contents as local heritage (for example, building a museum to a famous local author), even if there are limited financial benefits. In contrast, association with the contents is undesirable if they overwhelm the community (by obliterating other important aspects of local culture and identity) or the contents are deemed undesirable (communities may not want association with a drama about a serial killer, for example). So, local authorities (destinations) are aiming for levels of tourism that are profitable without being destructive/disruptive, and associations with contents that constitute a welcome component of community identity and pride. This is their doughnut.

Fans

For fans, the keywords of their doughnut are *relationships* and *experience*. Within their doughnut is a meaningful experience that enhances their relationships as fans with both the contents and the places/communities associated with the contents. An experience that falls short might be caused by attractions or communities that are deemed by fans to pay insufficient care or respect to the contents, for example, tacky commercialism. In contrast, a degenerative experience might be one in which the travel experience damages the fan's relationship with the contents themselves. For example, meeting an author at a book signing event and being treated very rudely might cause the fan to stop being a fan altogether. In this vein, changes in contents tourism caused by the naming and shaming of stars by the #MeToo movement might be an interesting topic of future research. However, defining what falls short of and what overshoots fans preferences depends very much on the values of the fandom. Disney fans have very different expectations to death metal fans, for example. However, Yamamura's research into anime tourism and community response has indicated that the most successful examples of contents tourism have occurred when a transition has taken place in fans from being a fan of the contents to a fan of the region, namely the places associated with the contents (Seaton *et al.*, 2017: 216–217). Based on this, 'getting into the doughnut' for fans is when the contents tourism experience enhances and deepens a fan's feelings of being a fan.

Contents businesses

The keywords of the doughnut for businesses are *reputation* and *copyrights*. As indicated at the beginning of this section, contents tourism is an unintended result of a mediatized work of popular culture. Contents businesses are not tourism operators. Revenues from tourism are not built into their business models when producing a set of contents. As such, contents businesses (which in our definition are professional producers of contents that range in scale from an individual author up to a multinational media conglomerate – see Seaton *et al.*, 2017: 23–26) are not necessarily interested in participating in contents tourism projects. Indeed, many eschew cooperation with tourism operators, particularly by refusing to allow the use of images, characters, or even the name of the franchise in tourist attractions. However, contents businesses may choose to participate if providing licensing rights is of significant commercial benefit. The other occasion on which the copyright holders will willingly participate is if doing so is an opportunity to strengthen loyalty among fans. If fans are able to purchase high-quality licensed goods while at tourist sites, meet creators or have other meaningful experiences, it may contribute to fan loyalty on the release of the next work in the franchise, or even a work in another narrative world controlled by the same copyright holder. Using the reverse logic, contents businesses deemed by their fans to fall short in the level of service provided to fans (by prohibiting official or limited-edition goods, for example) may find their fan loyalty reduced. Or, in their relationships with local authorities, contents business may find cooperation at the production stage (for example, location hunting) is less forthcoming next time if the business has gained a reputation for selfishly protecting its own interests regarding previous works. Conversely, overshoot for businesses means a cheapening of the brand, either through the injudicious distribution of licensing partners whose poor products diminish the brand, or simply by producing a glut of mass market goods that cheapen the contents.

Summary

These three intersecting doughnut rings, therefore, indicate that all three actors of contents tourism – local authorities, fans and contents businesses – have zones in which they may achieve their basic needs from contents tourism while not going beyond their ceilings into activities that cause degradation. By keeping contents tourism within these boundaries, the chances of limiting planetary degradation are also enhanced. Ultimately, it is in the congruence or healthy balance of the three actors' respective interests that truly sustainable contents tourism occurs. All cases studies of contents tourism failure may be thought of in terms of one of the actors falling outside of their doughnut. At this point, the relationship between the actors begins to break down and the tourism phenomenon declines. Fans may stop travelling to a site they consider does not cater to their needs as fans; local authorities may withdraw their welcome if there is overtourism or antisocial behaviour by fans; and businesses may withdraw licences if local authorities and fans disrespect their copyrights. If the watchword of the 21st century for human beings is living within our planetary means, then the watchword of contents tourism is a collaborative approach to the management of contents tourism in which the needs of all actors are satisfied, benefits are optimized and there is no overshoot into degradation either for oneself or others. If there is one ultimate key to achieving this happy balance, then as Yamamura (2011: 64) has argued, it is when the actors of contents tourism are all united in respect for and love of the contents that bring them all together.

Tourism in the Era of Global Contents

This book has focused on examples of international contents tourism in which people have travelled across national borders to engage more fully with narratives, characters, locations and other creative elements – 'the contents' – that they appreciate and love. In doing so, our research demonstrates the power of popular culture to bring people of different cultural backgrounds together via tourism. This process is substantially different to the many cultural interactions through tourism in which hosts perform their culture, tourists consume it and the ultimate focus is on cultural difference. Such interaction among travellers or between hosts and guests may be respectful, appreciative and even result in long-term friendships, but there is nothing specific between the travellers and hosts that binds them together beyond a commercial transaction. In contents tourism, in contrast, the focus is on the cultural togetherness among traveller-fans and with people in host communities (and sometimes contents business, too) forged by their shared and declared appreciation of a narrative world. Indeed, the contents are the reason they have come together. Of course, not every contents tourism encounter creates enduring or even amicable relationships. However, the potential for them is increased because travellers and host communities bonded by the contents can focus on what joins them rather than what divides them. In this book we have seen Taiwanese and Japanese connected by yōkai, Indonesian cosplayers who connect with other Asians via their joint love of Japanese pop culture, and gamers who met in Poland then visiting each other in their home countries. The example of 2.5-D theatre (Sugawa-Shimada, Chapter 8) has even shown that contents tourism can lead to 'contents migration' as fans seek a more long-term engagement with the mediatized culture they love. This is a phenomenon long recognized by people working in Japanese Studies departments, myself included, as we interview prospective students who admit they are interested in studying about Japan and in Japan because of a love of the popular culture, particular manga and anime.

This is the power of contents and contents tourism, with a vital supporting role played by the internet and social media, to break down barriers between cultures. In his seminal work on nationalism, Benedict Anderson highlighted the role of the media and 'print-capitalism' (Anderson, 1991: 36) in binding people together in the 'imagined community' of the nation. The 'imagination' is a vital concept within tourism studies, too, particularly in relation to 'places of the imagination' (Reijnders, 2011) or tourism 'imaginaries' (Chronis, 2012), and in a manner similar to the media's role in the creation of nations, the internet and social media have facilitated the creation of transnational imagined communities of fans. As a result, the process of contentsization – involving not only franchise extension by contents businesses but also fan productions, the activities of online communities and on-site tourist performances – is increasingly internationalized and globalized. Once the language barrier has been overcome, fans can congregate online with people half-way around the world just as easily as with people in the next street. International fandoms have emerged around contents ranging from Korean dramas to Jane Austen novels, and shared membership of a fandom becomes a powerful common language that may overcome many actual linguistic barriers.

Even the professional process of mediatized contents production is now thoroughly globalized. Many sets of contents are international or global from the outset, or they are deliberately expanded in ways to appeal to an international audience. Consequently, tourism induced by those contents is bound to be international from the outset. This is evident in two case studies that have barely featured in the published research emanating from our project thus far, despite being of key personal interest to the two project leaders and editors of this book – or, perhaps we have just avoided them (for now) because we would like to keep them as our own fan activities and separate from our professional lives as tourism researchers.

For Takayoshi Yamamura, watching the 1995 anime film *Ghost in the Shell* was a vital moment in developing his interest in anime culture and also China, where he subsequently spent a number of years doing research on heritage tourism. *Ghost in the Shell* was based on a Japanese manga, but the film directed by Mamoru Oshii was set in a futuristic city modelled on Hong Kong. When we visited Hong Kong together for a conference in 2017, Takayoshi pointed out the densely packed high-rise condominiums, which I immediately recognized when I watched the anime myself. The anime was released simultaneously in Japan, the US and UK – a sign of the emerging international fandom for futuristic anime started by *Akira*, which was released in 1988 (Japan), 1989 (US), and 1991 (UK). From a global contents perspective, the 2017 live-action remake starring Scarlett Johansson blended East and West, with most characters speaking in English and some in Japanese, while retaining the Hong-Kong-esque locations (albeit filmed mainly in Wellington, New Zealand!). From a global contents tourism perspective, however, Hong Kong is not the only city that claims to be the setting of the futuristic story. The manga and games are set in the fictional Niihama city or New Port City (for which there is no officially acknowledged model), but especially in two series of TV anime (the S.A.C. series and ARISE series, and some feature films associated with them), the city of Kobe and the surrounding area was the location model. Hence, Kobe has run various events and the city's tourism promotion division has collaborated with *Ghost in the Shell*. This creates an interesting and rare situation in which two cities in different countries have major claims to be the 'sacred site' for fans. Contents tourism divides between Kobe and Hong Kong, production is now divided between Japan and the USA, and the fandom is global. This is becoming increasingly common in the era of global contents.

Meanwhile, my own encounter with global contents has focused on the Harry Potter series. Early in the contents tourism project I had identified Harry Potter as a key case study. Before the films were produced there had been tourism by fans to Kings Cross Station in search of Platform 9¾, and then the films added shooting locations to the Harry Potter world that were not in the books, such as the Glenfinnan Viaduct in Scotland and the Millennium Bridge in London. Having watched (but not particularly enjoyed) *Harry Potter and the Philosopher's Stone* when the film came out in 2001, I had paid little attention to the unfolding global phenomenon. However, when I read *Philosopher's Stone* in 2016 in preparation for contents tourism fieldwork, I was quickly drawn into J.K. Rowling's magical world. As I worked my way through the series (I watched the film after finishing each of the seven novels) I was struck by the gradual internationalization of the Harry Potter world. The first novel was exclusively British, but references to other countries gradually increased. In the fourth novel, *Harry Potter and the Goblet of Fire*, there was the Quidditch World Cup and the Triwizard Tournament with Hogwarts competing against Durmstrang (Eastern Europe) and Beauxbatons Academy (France). By the final episode, the Battle of Hogwarts felt almost global in scale as the forces of good faced the forces of evil. After *Harry Potter and the Deathly Hallows*, the magical world then moved location to the USA for the *Fantastic Beasts* films. The franchise is now truly global. Despite there being no Japan connections in the novels/films (apart from one brief mention of a joke about a Japanese golfer in *Harry Potter and the Chamber of Secrets*), the magical world now extends to Japan. Worldwide Quidditch teams, including the Toyohashi Tengu, were described in *Quidditch Through the Ages* (a spin-off book written for the charity event Comic Relief), Japanese yōkai such as Kappa are counted among the Fantastic Beasts (Rowling, 2017: 48) and the announcement of a Japanese wizarding school (Mahoutokoro – from the Japanese for 'magic place') on the Pottermore website spawned fan art, YouTube videos and much speculation that there might be a Harry Potter anime in the pipeline. There is also a major phenomenon of Wrock (Wizard rock) Bands, which perform original songs about the Harry Potter world. The growth of Harry Potter, from a 1997 novel about a British schoolboy with an initial print run of 500 copies into a global cultural phenomenon is one of the largest ever contentsization processes. Likewise, the Harry Potter contents tourism phenomenon is global. In addition to the multitude of sites in the UK, there are the Wizarding World of Harry Potter attractions at Universal Studios in Orlando and

Figure 14.5 Global contents: Spiderman takes a photo of a cosplayer doing a kung-fu kick on the Hogwarts Express in Universal Studios Japan

Osaka (Figure 14.5), and countless unofficial or imitation cafes, events and conventions around the world where fans gather.

The multinational, multimedia and multiwork natures of *Ghost in the Shell* or Harry Potter and the global extent of their fandoms epitomize the reasons why contents tourism is a concept fit for analysing the connections between our globalized popular culture and global tourism activities in the digital, 21st century. The roots of contents tourism discourse may be in the niche world of Japanese anime tourism, as described at the beginning of the book, but in our research project as a whole we have attempted to take the concept beyond this niche level and to realize its full potential by applying it to the national, transnational and global levels. The ability of contents tourism to bring people from across the planet together is, for us, a positive and welcome opportunity in an age when confrontational identity politics and nationalism seem to be gaining ever more traction in national and international politics. At the same time, we recognize that too much of a good thing is undesirable. We strongly advocate a sustainable approach to contents tourism that encourages all actors to meet their basic needs while not overshooting to the point of detriment to and degradation of the environment. The 21st century challenge in contents tourism is to find, encourage and maintain that sustainable balance.

Note

(1) We have devoted relatively little attention to 'heritage and/or contents tourism' in this book, mainly because it was central to Seaton *et al.* (2017), and the special editions of *Japan Forum* (Volume 27.1, 2015) and *Journal of War & Culture Studies* (Volume 12.1, 2019).

References

Agarwal, S. and Shaw, G. (2018) *Heritage, Screen and Literary Tourism*. Bristol: Channel View Publications.
Anderson, B. (1991) *Imagined Communities*. London: Verso.
Beeton, S. (2016) *Film-Induced Tourism* (2nd edn). Bristol: Channel View Publications.
Beeton, S. and Seaton, P. (2018) Creating places and transferring culture: American theme parks in Japan. In S. Kim and S. Reijnders (eds) *Film Tourism in Asia: Evolution, Transformation and Trajectory* (pp. 251–267). Singapore: Springer.
Beeton, S., Yamamura, T. and Seaton, P. (2013) The mediatisation of culture: Japanese contents tourism and pop culture. In J. Lester and C. Scarles (eds) *Mediating the Tourist Experience: From Brochures to Virtual Encounters* (pp. 139–154). Farnham: Ashgate.
Chronis, A. (2012) Between place and story: Gettysburg as tourism imaginary. *Annals of Tourism Research* 39 (4), 1797–1816.
Frost, W. (2006) Braveheart-ed Ned Kelly: Historic films, heritage tourism and destination image. *Tourism Management* 27, 247–254.
Im, H.H. and Chon, K. (2008) An exploratory study of movie-induced tourism: A case of the movie *The Sound of Music* and its locations in Salzburg, Austria. *Journal of Travel & Tourism Marketing* 24 (2–3), 229–238.
JNTO (2019) Hōnichi gaikyakusū. See https://www.jnto.go.jp/jpn/statistics/since2003_visitor_arrivals.pdf (accessed January 2019).
JSPS (2014) International comparative research on the spreading and reception of culture through contents tourism. See https://kaken.nii.ac.jp/en/grant/KAKENHI-PROJECT-26243007/ (accessed January 2019).
Masubuchi, T. (2010) *Monogatari wo Tabi suru Hitobito. What is Contents Tourism?* Tokyo: Sairyusha.
Nanjangud, A. (2019) Bollywood tourism in Japan: Current challenges, potential directions. *International Journal of Contents Tourism* 4, 1–11.
Okamoto, T. (2013) *n-th Creation Tourism: Anime Seichi Junrei/Kontentsu Tsūrizumu/Kankō Shakaigaku no Kanōsei*. Ebetsu: Hokkaido Bokengeijutsu Shuppan.
Okamoto, T. (2015) Otaku tourism and the anime pilgrimage phenomenon. *Japan Forum* 27 (1), 12–36.
Raworth, K. (2017) *Doughnut Economics: Seven Ways to Think Like a 21st-Century Economist*. London: Random House.
Reijnders, S. (2011) *Places of the Imagination: Media, Tourism, Culture*. Farnham: Ashgate.
Rowling, J.K. (2017) *Fantastic Beasts and Where to Find Them*. London: Bloomsbury.
Scherer, E. and Thelen, T. (2017) On countryside roads to national identity: Japanese morning drama series (*asadora*) and contents tourism. *Japan Forum*. Published online 20 December 2017.
Seaton, P. (2015) Taiga dramas and tourism: Historical contents as sustainable tourist resources. *Japan Forum* 27 (1), 82–103.
Seaton, P. (2019a) On the trail of The Last Samurai (I): Taranaki. *International Journal of Contents Tourism* 4, 12–24.
Seaton, P. (2019b) On the trail of The Last Samurai (II): Hobbiton vs Uruti Valley. *International Journal of Contents Tourism* 4, 25–31.

Seaton, P. and Yamamura, T. (2015) Japanese popular culture and contents tourism: Introduction. *Japan Forum* 27 (1), 1–11.

Seaton, P., Yamamura, T., Sugawa-Shimada, A. and Jang, K. (2017) *Contents Tourism in Japan: Pilgrimages to 'Sacred Sites' of Popular Culture*. Amherst, NY: Cambria Press.

Watts, J. (2018) We have 12 years to limit climate change catastrophe, warns UN. *The Guardian*, 8 October. See https://www.theguardian.com/environment/2018/oct/08/global-warming-must-not-exceed-15c-warns-landmark-un-report (accessed January 2019).

Yamamura, T. (2011) *Anime, Manga de Chiiki Shinkō: Machi no Fan wo Umu Kontentsu Tsūrizumu Kaihatsuhō*. Tokyo: Tokyo Horei Shuppan.

Index

2.5-D theatre 13, 128–142, 238

actors (of contents tourism) 3–4, 43, 177, 193, 233–238, 241
actors (theatrical) 37, 125, 131–134, 138, 140, 142, 163, 167, 169, 193, 200, 210, 227, 229
  Burlinson, Tom 210–211
  Cranston, Bryan 197
  Firth, Colin 20, 26–28
  voice actors 129, 134, 139
adaptation (of works) 1, 7–10, 13, 19–22, 24, 27–32, 40, 42–43, 48–49, 62, 116, 129, 141, 162, 193, 225–226, 230
advertising 122, 177–178, 190, 226
America (see United States of America)
anime
  Evangelion 138, 180
  Ghost in the Shell 14, 239–241
  Heidi 13, 62–80, 230
  Kiniro Mosaic (see manga)
  layout system 64, 70, 80n
  Love Live! (see manga)
  Lucky Star 3–4
  NARUTO 129, 136, 138, 151
anime tourism/pilgrimage 2–4, 91–93, 128, 130, 170, 236, 241
artists xiii, 37–38, 40, 42, 49, 86, 154, 207
  Carigiet, Alois 68–69, 71, 77–78
  Segantini, Giovanni 69, 71, 77–79
  Twomey, Clare 40
augmented reality 130–131, 182
Australia 5, 14, 28, 141, 155, 167, 199–200, 205–214, 217
authenticity 13, 119, 126, 196–197, 202
authors/novelists 12–13, 19, 21, 26, 31, 34, 38–40, 46, 93, 95, 125, 193, 208, 226, 230, 236
  Austen, Jane 13, 19–32, 34, 42, 230–232, 239
  Brontës, the 13, 30, 34–43, 230, 232

  Roberts, Anna 93–94, 96
  Sapkowski, Andrzej 46–49
  Spyri, Johanna 63, 64, 66, 69, 71–72, 77
  Strayed, Cheryl 193–194
autoethnography 14, 189–190, 192–195, 207–208, 228

birthplaces 21, 24, 27–28, 35, 38, 79, 99
Bollywood 5, 29, 231
brand 134, 146, 176, 179, 182, 184, 186, 237
  brandscape 30, 34–43
  place brand 52

characters (see contents, definition of 'contents tourism')
  Harry Potter 14, 19, 30–31, 46, 89–91, 151, 240–241
  Hello Kitty 109, 150, 180–181, 183
  Pokémon 180–182, 184, 226
  Sherlock Holmes 8
China 8, 48, 120–122, 136–140, 150, 161, 163–164, 233, 239
Chinese fans 121, 137–141, 161, 164
climate change 232–233
consumers 7, 13, 19, 22, 36, 41–45, 48, 52, 58, 129–130, 132, 165–166, 171, 174, 190, 225, 228
contents
  contents brandscape 30, 35, 38–39, 41–43
  contents migration 137, 139, 238
  definition of 'contents' 4
  definition of 'contents tourism' 1–3, 6, 9, 11, 85, 174, 225–226
contents businesses 37, 225, 236–239
  CD Projekt Red 48, 50–53, 56
  Disney 181–183, 226, 231, 236
  Kadokawa 120, 131
  Studio Ghibli 64
  Zuiyo 63, 67, 72

contentsization 1, 7–13, 19, 23, 27, 31, 62, 70, 79–80, 116, 162, 170, 193, 195, 225–226, 230, 239–240
convergence 7, 19, 132
   convergence culture 10, 129
Cool Japan 180, 225
copyright 200, 226, 228, 230, 236–237
cosplay xiii, 3, 8, 10–11, 13–14, 23, 121, 129, 140, 144–156, 238, 241
   celebrity cosplayer 154
   hijab cosplay 148–149, 152
   Indocosu (Indonesian cosplay) 149
   industrial cosplay 185–186
   Jakarta Cosplay Parade 153
   World Cosplay Summit 151, 153
cross-referencing 68, 70–71, 78

directors 37, 167, 169, 191
   Gilligan, Vince 191, 196–197
   Kotabe, Yoichi 64, 67–70, 73, 77, 80–81n
   Miyazaki, Hayao 64, 66–70, 77
   Takahata, Isao 64, 66–70, 72–74, 77
   Tomaszkiewicz, Konrad 49–50
Disney (see contents businesses)
doughnut economics 232–237

economic impact 169, 189, 226, 235

fairy tales 50, 88–90
fandom 25, 27, 57, 116–119, 122, 125–126, 133, 141–142, 185–186, 230, 236, 239–241
fantasy 26, 46, 48, 55, 58, 66, 70, 86, 93–94, 110–112, 129–131, 171, 193
festivals 38, 111, 122, 189–190
   Anime Festival Asia 152
   cosplay festivals 145, 150, 152
   Jane Austen Festival Bath 19, 22–24, 31
   Man from Snowy River Bush Festival 210–213
   Osaka Asian Film Festival 167
   Shanghai International Film Festival 168
   Yōkai Matsuri Festival 104–108
film-induced tourism xiii–xiv, 5–6, 19, 22, 37, 46, 90, 150, 189–190, 192–193, 202–203, 213, 224–225, 227, 231

films
   Austenland 27
   Hello Stranger 27, 162, 166–171
   Lord of the Rings 19, 46, 227–228
   Man from Snowy River, The 210–211
   Star Wars 11, 174, 182–186
   Sound of Music, The 230–231
   War Horse 90
   Wild 194
   Witcher films, the 47–48
   Wuthering Heights 40
franchise 19, 31, 46–48, 51–55, 57–59, 120, 129–130, 182, 230, 232, 237, 239

games 1, 3, 8–9, 13, 34, 46–59, 85, 90, 120–122, 124, 128–130, 132, 134–141, 146, 155, 193, 225–227, 238–239
   Assassin's Creed 46
   Kingdom Come 47
   LARP (live action role-playing game, The Witcher School) 46–47, 53–58
   Pokémon Go 184, 226
   Token Ranbu (see theatre)
   video-game-induced tourism 47–48
   Witcher 3: Wild Hunt, The 48–53
gaze (the 'tourist gaze') 13, 85–86, 119, 193
goods (see merchandize)
government 1–3, 6, 37, 52, 66, 85, 92, 100, 106, 108, 150, 153, 154, 156, 161–162, 169, 178, 180, 184, 224, 225, 227, 232
graves 21, 24, 26, 28, 35, 123, 212
guided/package tours 41, 225, 228–229
   Biking Bad tour 196–199
   Breaking Bad RV tour 199–202
   Hello Stranger tours, and tours in *Hello Stranger* 167–170
   Hidden Britain Tours 27, 31
   horseback safari 206, 210, 213
   JASNA (Jane Austen) tours 22, 26–27, 31
   Walk Japan 216, 218

Hallyu (see Korean Wave)
heritage xiii, 6, 30, 41, 52, 58, 111, 226, 228, 235, 239
   heritage and/or contents tourism 230

heritage (Continued)
  kai 'mystery' heritage 98–104, 110
  UNESCO World Heritage 218
historical contents/dramas 12, 46, 163, 230
  Dae Jang Geum 161, 166
  Romance of Three Kingdoms 8
historical figures 131, 137, 230
  Sanada, Yukimura 8
Hong Kong 5, 121–122, 136, 164, 239–240

idols/*aidoru* 2, 116, 119–120, 122, 139, 154, 165, 226
imaginary
  imaginary places 3, 66
  tourism imaginary 12–13, 19–20, 64, 70, 79, 238
imagination 8, 13, 19–20, 22, 43, 49, 69, 71, 78–79, 89, 118, 129, 164, 171, 190, 206, 212, 238
  religious imagination 116
internet 50, 52, 54–56, 58, 118–119, 131–132, 135–136, 139–140, 146, 169, 182, 225, 232, 238
Islam 122, 148–149

kawaii (cute) 98, 101, 109, 111–112, 120, 139, 180
kontentsu (contents) xiv, 1, 35, 130
Korea, Republic of 5, 7, 13, 116, 119–125, 135–136, 155, 161–171, 239
Korean Wave 161–168, 171
Koreanness 13, 162–166, 168–171

literary tourism 19–22, 25, 27, 31–32, 34, 37–38, 40–41, 64, 194, 206–207, 225, 230, 232
literature (see also novels) 2–3, 8, 12, 19, 28, 38, 41, 48–51, 62–64, 73, 117, 193, 205–206, 208–209, 230
  children's literature 86, 88–90, 93, 95–96
  oral literature 112n
livery 175–181, 184–186
local authorities 42, 225, 227, 229, 235–237
locations
  film location tourism 46, 165, 168, 190, 193, 198–199, 228–229, 240
  filming locations 24, 26–27, 30, 43n, 165–167, 189–190, 227–228, 240

location hunting 21, 64, 66–73, 79, 81n, 227, 231, 237
logo 36, 177–179, 181, 184
Low Cost Carriers (LCC) 161, 179

managers of tourist sites (interviews)
  Cooper, Caron 92, 96
  Haruyama, Marie 94–96
  Kusdianto, Tossy 151
  Lin, Chih-Ying 108
  Nagamoto, Kazuaki 113n
  Shōji, Yukio 99
  Smith, Madelaine 25
manga 1, 3, 7, 9, 13, 34, 85, 90–91, 97n, 99, 102, 104, 111, 128–130, 132–136, 139, 141, 144, 148–150, 152, 165, 193, 225, 227, 231, 238–239
  GeGeGe no Kitarō 99–100
  Kin-iro Mosaic 90–93
  Love Live! 116, 119–126
manga artists
  Hara, Yui 91–92
  Mizuki, Shigeru 99, 100, 102, 110–111
  Yamazaki, Kore 97n
media mix 5–7, 9, 25, 65, 116, 119, 129–132, 134, 137, 141, 230
mediatization 5, 63
merchandise 52, 58, 92, 121, 123–125, 134, 137, 139, 141, 155–156, 183, 186, 229, 237
motivation 20, 26–27, 29, 31, 57–58, 150, 161, 167, 176–177, 190, 208, 231
multi-use 5–9, 12, 19, 37, 47, 62
museums xiii, 51, 216, 226, 236
  Brontë Parsonage Museum 40, 42
  Hyōgo Prefectural Museum of History 98
  Jane Austen's House Museum 21–22, 24–25, 27, 31
  Johanna Spyri Museum 66
  London Film Museum 232
  Man from Snowy River Museum 212
  Mizuki Shigeru Museum 99–100
  Segantini Museum 69, 77
  Yōkai-yashiki (Yōkai House and Stone Museum) 104–105, 107, 112

music 35, 64, 67, 69, 109, 117, 120, 139, 161, 163–165, 181, 206, 213–214, 222
  concerts 29, 120, 135, 137
  musicals (see theatre)
  musicians 37
  music-induced tourism 37
  Sound of Music, The 230
myth/mythology 13, 46, 49–50, 56, 58, 117, 134, 230

narrative
  narrative quality 1, 3, 5, 12, 30–31, 85, 94, 206, 229
  narrative world 1, 3–4, 6–13, 19, 27, 30–32, 34–35, 37–38, 42–43, 46, 48, 54–58, 62–66, 68, 70–72, 79, 116, 130, 134–135, 141, 162, 170, 193–195, 225–226, 230, 237
  tourism narrative 208
New Zealand 19, 46, 227–228, 231, 239
novels 1, 7–8, 13, 19–22, 24–31, 34–35, 38, 40, 42, 43n, 46–51, 55, 63, 71, 74, 85, 90, 120, 128, 131, 139, 150, 193, 225–227, 230, 239–240
  Austenland 27, 31
  graphic novels 19
  Harry Potter novels 240
  Heidi (see anime, *Heidi*)
  light novels 130
  novelists (see also authors) 12
  Pride and Prejudice 20, 22, 24, 26–29, 31, 91
  Tale of Genji, The 8
  Wild 193
  Witcher, The 47–50, 55

otaku 3, 92, 123, 149

performance 2, 12–13, 19, 23–24, 112n, 118, 120–121, 123, 125, 128–130, 132, 134–142, 148, 154, 167, 195, 231, 239
  performance turn 13
photos/photography 11, 26, 49, 68–70, 91–92, 96n, 119, 122–123, 136–137, 141, 146, 155, 177, 180, 182, 185, 195–196, 198, 200–202, 212, 241
pilgrimage 1–4, 6, 12–14, 21, 26–27, 31, 40, 46, 86, 92–93, 116–120, 122–123, 125–126, 170, 186, 189–190, 192–196, 202, 214–216, 219, 222

seichi junrei (sacred site pilgrimage) 128, 141
planes 13–14, 174–186
  All Nippon Airways (ANA) 174, 178–186
  Japan Airlines (JAL) 177–182
poetry 42, 205–222
  haiku 206–207, 214–215, 217, 219–221
  Man from Snowy River, The (A.B. 'Banjo' Paterson) 205–207, 209–212
  Narrow Road to the Interior (Basho Matsuo) 205, 214, 216–217, 221–222
Poland 46–59, 192, 238
policy 1–4, 14, 224, 226–227, 232
producers (profession) 19, 169, 227, 236
  Nakajima, Junzo 64, 67–68, 70, 80n
  Takahashi, Shigeto 64, 67, 72, 76
prosumers 11, 19

reality (see also augmented reality, virtual reality) 53, 56, 64, 66, 68, 89, 96, 109, 120, 129–132, 141, 163, 171, 176, 194, 203, 227, 232
religion 116–117, 119, 126, 148–149, 222
researchers
  Academy of Contents Tourism 2–3, 224
  Beeton, Sue 5–6
  Japanese Association for Contents History Studies 224–225
  Jenkins, Henry 6–7
  Masubuchi, Toshiyuki 2–3, 224
  Okamoto, Takeshi 4, 130, 224
  Raworth, Kate 233–235
  Reijnders, Stijn 2–3, 206, 238
  Shimooka, Shōichi (Dr Yōkai) 102, 104–107
  Tzanelli, Rodanthi 6
  Yanagita, Kunio 99, 101–102
ritual 12–13, 59n, 118–122, 124–126

sacredness 12–13
  sacred sites 1, 3, 4, 12–13, 116, 118, 125, 128, 130, 136, 141–142, 186, 240
  sacred space 138
  sacred spots 136–137, 141–142
settings 22, 24, 34–35, 37–38, 42, 49, 53, 58, 93, 97n, 100, 137, 148, 193, 239
shrines 21, 92, 109, 118, 219

singers
  Amuro, Namie 181
  Arashi 181–182
sites of contents tourism (attractions)
  airports 175–177, 184
  Alm hut 67, 69, 71–73
  Dreamton 90, 94–95
  Fosse Farmhouse 91–93
  Heididorf 64–65, 73, 78
  Hobbiton 228–229
  King's Cross Station 30, 240
  Lyme Park (Pemberley) 21–24, 30–31
  Mount Umi 122–126
  Xitou Monster Village 98, 108–112
  Yamashiro Ōboke Yōkai Village 98, 104, 107–108, 110
sites of contents tourism (towns, regions)
  Albuquerque 189–192, 195–197, 199–200, 203
  Bandung 144–146, 148–152, 156
  Bath 19, 22–24, 30–31
  bush (Australian) 205–210, 212–213
  Castle Combe (Cotswolds) 5, 85–96, 231
  Gdańsk 49–50
  Haworth 34–35, 38–40, 42–43
  Maienfeld 65, 67–69, 72, 74, 78
  Sakaiminato 99–101, 106, 110–111
  Seoul 121–122, 167–170
  Tōno 99–103, 106, 110–112
  Winchester 21, 24, 26, 30
  Yamashiro 98–99, 101–102, 104–108, 110–112
social media 27, 51, 54, 116, 118–119, 121–123, 126, 130–132, 134, 137, 139–141, 146, 148, 177, 190, 195, 201, 206, 217, 222, 233, 238–239
  Facebook 54, 57, 136, 146, 169
  Instagram 119, 146
  Niconico dōga 132–133
  Snapchat 190, 195
  Twitter 119, 125, 131–132, 136–137, 141–142
  YouTube 52, 132–133, 141, 148, 169, 182, 222, 240
societies
  Brontë Society 40
  Jane Austen Society of North America 22, 24
solo travel 27, 190

spin-off works 19, 24, 26, 31, 203, 214, 240
spiritual meaning 27, 215, 217
storytelling 7, 9, 61, 129, 190
  transmedia storytelling (see transmedia)
sustainability 4, 203, 233, 235
Switzerland 63–70, 73–74, 77, 79

Taiwan 13, 98–99, 108–112
television 1–3, 7–8, 20, 22, 34–35, 48, 52, 55, 62–64, 66, 69, 74, 85, 88, 90–91, 99, 122, 128–129, 131–132, 134, 137, 142n, 144, 148, 154, 161, 163, 165, 175, 189–190, 192–193, 196, 198, 203, 206, 209–210, 212, 222, 225, 229, 239
television anime 62, 64, 66, 69, 81n, 129, 131–132, 134, 137, 142n, 239
television dramas 1–3, 7–8, 34, 85, 90, 128, 144, 161, 163, 165, 175, 189, 190, 193, 210, 225, 229
  Breaking Bad 189–192, 194–203
  Dae Jang Geum 161, 166
  Man from Snowy River, The 210–211
  Morning Dramas (Japan) 229
  Pride and Prejudice 20, 22, 24, 26, 28–29, 31, 91
  Taiga Dramas (Japan) 229
  Winter Sonata 161, 168
Thailand 13, 161–171
theatre 8, 13, 19, 131–142, 152, 238
  movie theatres 121–122, 167
  Musical Prince of Tennis 130, 133–138
  musicals 22, 24, 63, 90, 128, 142
  Rose of Versailles, The 134
  Takarazuka Revue 63, 128, 134, 142
  theatre tourism 128, 130, 134–142
  Token Ranbu 130, 135–138, 140
theme parks 13, 95, 98, 109, 154, 229, 231
  Dae Jang Geum theme park 166
  Tokyo Disneyland 181–183, 226, 231
  Universal Studios Japan 231, 240–241
toys 89, 93–94, 121
trains 43n, 87, 122, 176–177, 180–181
transmedia 1, 5, 7–12, 63, 79, 129, 230
  transmedia storytelling 7, 9, 129
transnational 1, 5, 9–12, 14, 62–64, 79, 161, 165, 174–175, 181, 239, 241

United Kingdom 13, 19–32, 34–43, 85–96, 199
United States of America 11, 20, 19–32, 54, 129, 150, 181, 189–203, 224, 231
universities 108, 124, 139, 141, 145, 151–152, 213

virtual 120, 129–131
  virtual reality 130

worlds (see narrative world)
wrapping culture 13, 175–176, 180, 185–186
  Hello Kitty shinkansen 180–181, 183
  itasha (wrapped cars) 180

yōkai 13, 98–112, 230, 238, 240
  Kappa 100–101, 111–112, 240
  Konaki-jiji 99, 102, 104, 106, 111

For Product Safety Concerns and Information please contact our EU Authorised Representative:

Easy Access System Europe

Mustamäe tee 50

10621 Tallinn

Estonia

gpsr.requests@easproject.com

www.ingramcontent.com/pod-product-compliance
Ingram Content Group UK Ltd.
Pitfield, Milton Keynes, MK11 3LW, UK
UKHW020240200525
458704UK00018B/129